Reviews of Physiology, Biochemistry and Pharmacology 132

Springer-Verlag Berlin Heidelberg GmbH

Reviews of

132 Physiology Biochemistry and Pharmacology

Editors
M.P. Blaustein, Baltimore   R. Greger, Freiburg
H. Grunicke, Innsbruck   R. Jahn, Göttingen
L.M. Mendell, Stony Brook   A.Miyajima, Tokyo
D. Pette, Konstanz   G. Schultz, Berlin
M. Schweiger, Berlin

Honorary Editor:
E. Habermann, Gießen

With 20 Figures and 4 Tables

Springer

ISSN 0303-4240
ISBN 978-3-662-31001-4   ISBN 978-3-540-69581-3 (eBook)
DOI 10.1007/978-3-540-69581-3
Library of Congress-Catalog-Card Number 74-3674

Originally published by Springer-Verlag Berlin Heidelberg New York in 1998.

Softcover reprint of the hardcover 1st edition 1998

Production: PRO EDIT GmbH, D-69126 Heidelberg
SPIN: 10551540   27/3136-5 4 3 2 1 0   – Printed on acid-free paper

Contents

Indexed in Current Contents

Mechanosensitive Ion Channels in Nonspecialized Cells[*]

F. Sachs[1] and C.E. Morris[2]

[1]Biophysical Sciences, State University of New York, 120 Cary Hall, Buffalo, N.Y. 14214, USA
[2]Neurosciences, Loeb Institute, Ottawa Civic Hospital, 1053 Carling Ave, Ottawa, Ontario K1Y 4E9, Canada

Contents

---

[*]We dedicate this review to Dr. Harold Lecar, our insightful teacher

1
Introduction

Mechanosensitivity is a universal property of cells and provides an obvious survival advantage for free living organisms who must back away from barriers and evade predators. For multicellular animals, mechanosensitivity provides sensory inputs for hearing, touch, kinesthesis and proprioception, the volume of hollow organs, the regulation of hormones involved with fluid homeostasis and the regulation of growth of bone and muscle. In plants, mechanosensitivity ensures that roots grow down and stems grow up, that pea plants twine and that the gills of fungi are vertical.

The transducers for mechanical inputs are not clearly characterized, but must include the cytoskeleton and its mechanochemical components such as the actin and tubulin based transporters, cell surface proteins such as the integrins, and ion channels. Of all the mechanical transducers, mechanosensitive ion channels (MSCs) are the simplest to study since their response is rapid and the establishment of cause and effect is a simpler process than most. In contradiction to established approaches in biology, the most valuable sources of information about the properties of these channels has not been obtained from specialized cells but rather from cells whose primary function does not appear to be sensory. This is because the specialized transducers are firmly bonded to the structures whose strain they are to sense and hence difficult to isolate without damage. A consequence of this rather bizarre choice of cells is that the physiological function of the channels is often in question.

The designation of a channel as mechanosensitive is empirical and signifies only that the channel's open probability responds to membrane deformation. MSCs may be either stretch-activated channels (SACs) or stretch-inactivated channels (SICs). Some SACs appear to be curvature sensitive, only responding when bowed toward the nucleus; examples are found in glial cells (Bowman et al. 1992; Bowman and Lohr 1996), smooth muscle cells (Kawahara 1993) and endothelial cells (Marchenko and Sage 1996). Other than mechanical sensitivity, the phenotypes are quite varied, suggesting that they do not all belong to a single homologous family of channels. Some are activated by stimuli other than mechanical stress, while others appear to be novel. It is almost certain that MSC gating is a response to tension conveyed to the channel via either the surrounding lipids or the membrane cytoskeletal elements. Although physiological tasks requiring mechanosensors can be pointed to in most, if not all, cells in which MSCs have been found, it has been frustratingly difficult to prove that these tasks are carried out by the MSCs.

Over the last few years, this impasse has started to give way and several exciting advances have occurred: an MSC channel from bacteria has been cloned, an excellent candidate for a component of an MSC has been cloned from touch sensitive mutants of *Caenorhabditis elegans,* and MSC activity has been linked to a well-characterised physiological function in an osmosensory neuron. In addition, there is an increasing realisation that the state of the cytoskeleton adjacent to the membrane may strongly affect the response of MSCs to mechanical stimuli. Taken together, these developments provide a good reason for reviewing the state of the field. We do not attempt to offer a compendium of recent findings since several reviews of recent vintage are available (Sackin 1995; Martinac 1993; Lecar and Morris 1993; Morris and Horn 1991a; Morris 1995; French 1992; Yang and Sachs 1990; Hamill and McBride 1995; Bargmann 1994; Sukharev et al. 1997).

1.1
Scope of This Review

We start with an operational definition of MSCs then move to a consideration of mechanical forces in the context of membranes and ion channels. After summarizing a general stochastic model for two-state MSCs, we discuss several physical models of MSC gating, giving particular emphasis to bilayer mechanics. Since the behavior of MSCs varies markedly with the method of study, we devote considerable space to techniques for recording and stimulating. Since eucaryotic MSCs appear to require the membrane skeleton for activity, we treat the theory and evidence relating to its role. Transient responses of MSCs (adaptation, delay) are covered in this context and also discussed in their own right. We include a section on the pharmacology of MSCs and a section on the role of MSCs in cell and organ physiology.

1.2
What Are MSCs?

MSCs are defined by the distinctive property that their gating is responsive to membrane deformation. It is reasonable to ask what is meant by "responsive," since a number of papers have been published in which mechanical stimulation caused only slight changes in kinetics. For example, the NMDA channel has a mechanosensitive dynamic range of ≈3 (Paoletti and Ascher 1994), and the bovine epithelial Na⁺ channel ENaC has a dynamic range (Awayda et al. 1995). In contrast, truly mechanosensitive channels are specialized in the sense that they can vary by orders of magnitude their probability of being open as a function of mechanical stress. Modulation of gating by mechanical stress does not per se justify classifying a channel as "mechanosensitive" any more than modulation of a nicotinic AChR channel by voltage (Magleby and Stevens 1972) classifies it as a "voltage sensitive" channel. Although these distinctions are somewhat arbitrary, they help to define the interim taxonomy of channels. Eventually we would hope to apply the designation "mechanosensitive" to channels that recognize mechanical deformation as a meaningful physiological signal. Under these conditions, a dynamic range as small as 2–3 could be significant. Until then, it would be wise to use a term such as "weakly mechanosensitive" to describe channels with a small dynamic range.

1.3
Activating MSCs

The requisite property of MSCs is that the channel's dimensions must change during the rate limiting transitions. If the open states are larger than the closed states, the channel is a SAC, and the opposite for SICs. In the case of a SAC, for example, the applied force does work on the channel so that the open state is lower in energy than the closed state under the same force, and has a favored occupancy.

The activation is allosteric and stochastic in the same sense as the gating of other channels. Pulling on the channel does not change its active site, i.e., its conducting pore; rather, pulling changes the probability that the conducting pore will be open. In an exception to this generalization, one report claims that the reconstituted epithelial Na⁺ channel, αβENaC, loses its intra-cation selectivity with deformation of the artificial bilayer (Awayda et al. 1995). In eucaryotes, forces appear to be transferred to the channels by the cytoskeleton rather than by the lipids. A functional equivalent of the model of Fig. 1 is shown in Fig. 2 with cytoskeletal attachments. Although Fig. 2 is drawn with the channel pulled in one direction, in reality, the channel may be pulled in several directions to produce a discrete form of tension.

Fig. 1. Cartoon of an allosteric gating mechanisms for SACs and SICs. In each case the larger diameter conformation of the channel protein should become more stable with increased in-plane tension. The open channel conductance does *not* change with tension, nor is it expected to with low levels of hydrostatic pressure (Sachs and Feng 1993; Rosenberg and Finkelstein 1978). Note that both conformations occur with some probability under *all* tensions. Tension is envisaged as pulling on the channels' perimeter and favoring one or other conformation

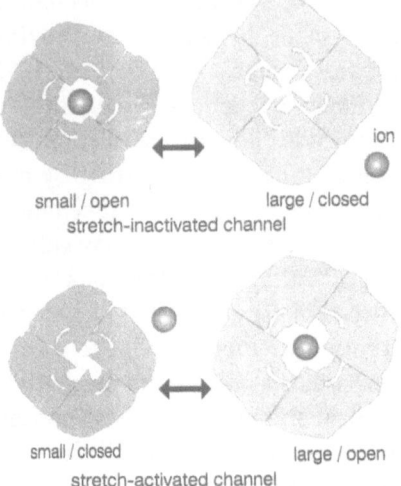

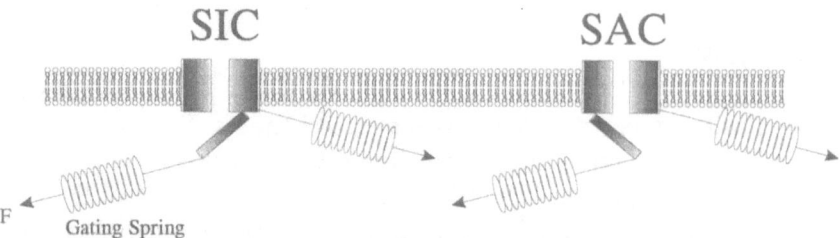

Fig. 2. Cartoon of the two types of MSCs in which force is transferred to the gates via the cytoskeleton rather than the lipids. *F*, Force within the attached cytoskeleton and the "gating spring" represents the elasticity of these elements. The channel gates should be viewed as bistable elements that are either open or closed, snapping between the two states with a probability controlled by the force F

In the absence of structural data, MSCs must be defined phenotypically by the way they respond to stimuli, and one cannot separate the present definition of the channels from the method by which they are stimulated. Additionally, in the absence of specific pharmacological reagents for MSCs, whole-cell mechanosensitive currents cannot be clearly identified as arising from MSCs. Such currents may arise via activation through second messengers produced by other mechanosensitive enzymes (Chen et al. 1995; Lehtonen and Kinnunen 1995; Watson 1990). It is generally not

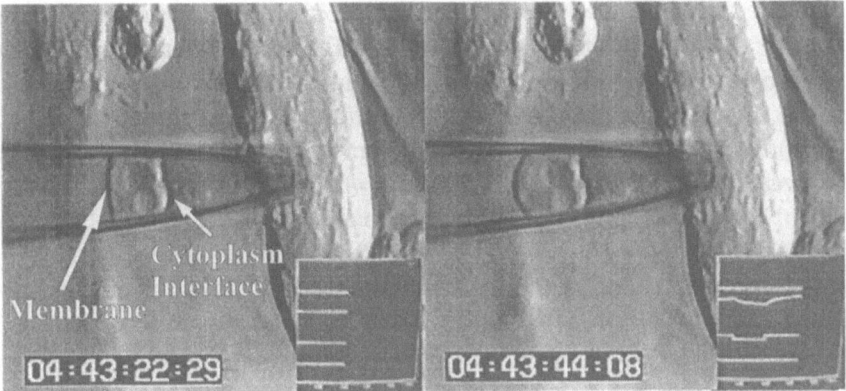

Fig. 3. Images of a cell-attached patch from an *mdx* myotube in culture. As is common with many patches, a plug of membrane-covered cytoplasm has pulled away from the cell leaving a saline filled gap between the plug and the remaining cellular cytoplasm. *Left*, with zero applied pressure; *right*, with –40 mmHg. With increased suction, the membrane curvature increased. The excised plug of cytoplasm was intact as evidenced by the lack of Brownian motion of organelles. When the cytoplasm degenerates, Brownian motion visibly increases. Note the curvature at zero pressure illustrating the presence of residual stress in the cytoplasm

simple to identify a whole-cell current that arises after mechanical stimulation as one arising from a direct action on MSCs. The potential difficulties involved in the study of MSCs is shown by the complexity of patch structure (Sokabe et al. 1991), an example of which is illustrated in Fig. 3.

2
Forces

To understand the activity of MSCs, it is essential to understand the forces involved and methods for generating them. Since the units used in cell mechanics are unfamiliar to most electrophysiologists, we have included tables of common units of pressure and tension. Because the literature contains CGS, MKS and other miscellaneous units, our text is also mixed. For pressure, we try to stick with the mmHg which is common to daily weather pressure reports and is intuitively satisfying. For tension, we tend to use the CGS dyne/cm which is the same as the molecular scale unit, pN/nm. One other definition to be declared and apologized for at the outset is T, used for both tension and temperature. As temperature, T only appears as $k_B T$, Boltzmann's constant times absolute temperature. To assist in comparing data using different units we have created the conversion tables for pressure (Table 1) and tension (Table 2). For those unfamiliar with mechanical units some "rule-of thumb" MKS values may be useful: the weight of a cell's nucleus (in water) is ≈ 0.2 pN; the actin-gelsolin bond ≈ 20 pN; dissociation of actin from myosin and the power stroke of myosin ≈ 9 pN; the actin-actin bond ≈ 110 pN (Nishizaka et al. 1995) the biotin-streptavidin bond ≈ 250 pN (Grubmuller et al. 1996); a carbon-carbon bond $\approx 30\,000$ pN (Berendsen 1996). For pleasant historical flavor, the reader is reminded that an apple weighs approximately 1 Newton (≈ 100 g).

In attempting to place an order of magnitude value on the forces involved in changing the state of an MSC, we could describe the mean force, F, on a channel by multiplying its perimeter by the tension. A typical tension for half activation of an MSC of 1 dyne/cm gives, for a 5-nm diameter channel, $F = 2\pi r T = 15.7$ pN. It has been difficult to measure the forces that activate MSCs in cell membranes since, in general, they consist of parallel and series components including the extracellular matrix, the bilayer and the cytoskeleton. Experiments on patches (Sokabe et al. 1991) and with "whole-cell" recording from vesicles (Gustin et al. 1988) suggest that MSCs can be half activated with mean tensions in the range of 1 dyne/cm. Some, however, such as the recently cloned bacterial MSC, denoted MscL, may require much more. Since some MSCs have recently been reconstituted in lipid bilayers (Sukharev et al. 1997; Hase et al. 1995; Awayda et al. 1995), it should be possible to form more precise estimates of tension dependence of activation in these systems, but this has not yet been done. A cartoon of a closed-open transition of an MSC in a lipid bilayer is shown in Fig. 4. Tension decreases the lipid density and favors the larger conformation of the channel. It should be pointed out that in eucaryotic cells, stresses need not be only in the plane of the membrane since the cytoskeleton (Sokabe and Sachs 1990), and perhaps the channels themselves, can exert forces normal to the membrane.

Table 1. Pressure in various units

	N/m^2(Pa)	$dyne/cm^2$	cmH_2O	mmHg	atm	mbar	k_BT/nm^3
N/m^2(Pa)	1	1.00×10^1	1.02×10^{-2}	7.50×10^{-3}	9.86×10^{-6}	1.00×10^{-2}	2.47×10^{-7}
$dyne/cm^2$	1.00×10^{-1}	1	1.02×10^{-3}	7.50×10^{-4}	9.87×10^{-7}	1.00×10^{-3}	2.47×10^{-8}
cmH_2O	9.80×10^1	9.80×10^2	1	7.35×10^{-1}	9.67×10^{-4}	9.80×10^{-1}	2.43×10^{-5}
mmHg	1.33×10^2	1.33×10^3	1.36×10^0	1	1.32×10^{-3}	1.33×10^0	3.30×10^{-5}
atm	1.01×10^5	1.01×10^6	1.03×10^3	7.60×10^2	1	1.01×10^3	2.51×10^{-2}
mbar	1.00×10^2	10^3	1.02×10^0	7.50×10^{-1}	9.87×10^{-4}	1	2.47×10^{-5}
k_BT/nm^3	4.05×10^6	4.05×10^7	4.12×10^4	3.03×10^4	3.99×10^1	4.05×10^4	1

Multiply the left unit by the factor to obtain the upper unit.

Table 2. Tension in various units

	N/m	dyne/cm	pN/nm	erg/cm^2	kBT/nm^2
N/m	1	10^3	10^3	10^3	2.47×10^2
dyne/cm	10^{-3}	1	1	1	2.47×10^{-1}
pN/nm	10^{-3}	1	1	1	2.47×10^{-1}
erg/cm2	10^{-3}	1	1	1	2.47×10^{-1}
k$_B$T/nm^3	4.05×10^{-3}	4.05	4.05	4.05	1

Multiply the left unit by the factor to obtain the upper unit.

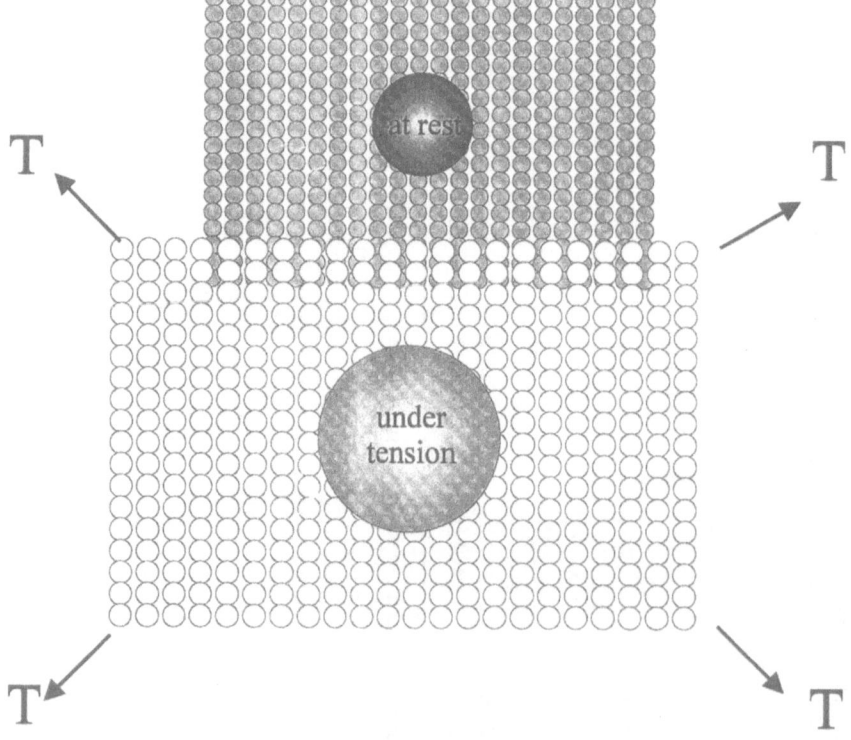

Fig. 4. Cartoon of a SA channel in a bilayer under tension (T). The open channel (not necessarily the permeation pathway) is larger than the shut channel so that under tension T, an amount of work equal to TΔA is done on the channel when it passes from closed to open. ΔA in this case is $8p(r_o^2 - r_s^2)$ where r_o is the radius of the open channel and r_s is the radius of the shut channel. Under tension the bilayer lattice (*shaded*) expands (*clear*) and is shown greatly exaggerated for clarity. The lipid area expansion is usually < 3%

2.1
Bilayers

To help define the minimal elements of membrane mechanics necessary to think about MSCs, we first consider the bilayer. The bilayer is a two dimensional fluid characterized by three intrinsic mechanical parameters (see (Bloom et al. 1991) for an excellent and thorough review on the physical properties of bilayers).

2.1.1
Changes in Area

The first property is the resistance to increases in area, or stretching, characterized by the elastic constant K_A. This is defined by $T = \Delta A/A\ K_A$, where A is the original area and Δ A is the increase in area under tension T. A bilayer has relatively high resistance to being stretched ($K_A \approx 100$–1000 dyne/cm) and lyses with expansions of 3%–5% depending on composition.

2.1.2
Changes in Bending

The second property is the resistance to bending and is defined through the bending moment, $M = k_c \times \Delta(1/r_1 + 1/r_2)$ where the r's are the principal radii of curvature and k_c is the bending rigidity. A bilayer bends easily ($k_c \approx 10^{-12}$ dyne-cm) unless the tension is high. Because of the low bending rigidity, an unstressed membrane under the influence of thermal energy flaps in a direction normal to the plane of the membrane. Such "drum head" modes of oscillation can be seen in red cells at normal volume where the membrane is flaccid (Gallez and Coakley 1986). These surface waves act as entropic "springs" that decrease, by about 20%, the theoretical elasticity that is associated with decreases in density under tension (Bloom et al. 1991). The bending fluctuations are always present and their amplitudes decrease with the mean tension. At tensions greater than 10^{-3} dyne/cm, the fluctuations are smaller than the wavelength of light and hence are generally unobservable.

2.1.3
Changes in Shape

The third constitutive property of bilayers is the shear rigidity, the resistance to changes in shape at constant area. As the bilayer is a fluid, shear rigidity is zero at equilibrium but the corresponding shear viscosity may be significant during membrane flow.

2.1.4
Activation of Channels by Tension

In the simple view of an MSC as a bistable, thermally activated object; activation requires that the channel's energy exceed the barrier(s) separating open from closed states. The channel is not *dragged* open by the applied force (Sachs and Lecar 1991; Lecar and Morris 1993). The probability of being open is affected by the applied force because the depth of the energy wells corresponding to the closed and open states depend on the applied force according the Boltzmann distribution (Hille 1992). The work done on the channel's sensor domain by tension (force) would be on the order of $T\Delta A$ where ΔA is the difference of in-plane area between the closed and the open channel. A stretch *activated* channel needs to have a larger dimension of the sensor when the channel is open than when it is closed. The change in dimensions can be estimated from an analysis of the gating process.

Consider a two state channel, $C \leftrightarrow O$ (closed $\leftrightarrow$ open). In the absence of tension, the probability of being closed or open is, respectively, $P_c(0)$ and open, $P_o(0)$. In the presence of a tension, T, these probabilities become $P_c(T)$ and $P_o(T)$. From the Boltzmann relationship:

$$P_o(T)/P_c(T) = e^{-\Delta G(T)/kBT} \tag{1}$$

where $\Delta G(T)$ is the free energy difference $G_o - G_c$ between O and C at tension T. For our two-state channel, $P_c = 1 - P_o$, and:

$$P_o(T) = 1/(1 + e^{\Delta G(T)/kBT})$$

The simplest model for $\Delta G(T)$ is Hooke's Law for area, $\Delta G(T) = \Delta G(0) - T\Delta A = (T_0 - T)\Delta A$, where $\Delta G(0)$ represents the energy difference between open and closed states at 0 tension. It can be rewritten as an offset tension,

T_0, times the change in area between the open and closed states, ΔA. Thus, we can rewrite Eq. 1 as:

$$P_o(T) = 1/(1+e^{\Delta A(T-T0)/kBT}) \tag{2}$$

and this is plotted in Fig. 5. The maximal slope, which occurs at the inflection point at T_0, is equal to $\Delta A/4k_BT$. If P_o is measured as a function of tension, the maximal slope can then be used to infer ΔA, the change in area between the closed and open states.

A minimal estimate of the true slope is obtained from knowing P_o at two different tensions (e.g., T_1 and T_2). As shown by the slope of the straight line in Fig. 5, this is always less than the true slope, but can be used to place a lower limit on the free energy that can be added to a channel. With a little algebra, Eq. 1 can be solved for ΔG in the presence of two different stimuli, noted 1 and 2. For the two state system the free energy difference between the open states is $\Delta G_{12} = k_BT \ln[(1/P_o(2)-1)/(1/P_o(1)-1)]$. This function is plotted in Fig. 6.

If the probability of being open at rest is 10^{-3}, and the probability of being open with stimulation approaches 1, then ΔG_o7 kT. This number is a minimal estimate (it ignores inactivation or adaptation) and simply corresponds to the equilibrium populations of the open and closed states in a thermal bath. Thus, one need not do a detailed dose-response curve to make an estimate of the free energy input to the gating process. This is important because, as pointed out below, the observed slope of the P_o curve may vary considerably, based upon how much of the stress is borne by parallel elastic elements, and one may falsely assume that a channel with a low slope is intrinsically insensitive to stress. The minimal dynamic range of the channel, $P_o(T1)/P_o(T2)$, provides a minimum estimate of ΔG without knowledge of the influence of the elastic elements. If the mean tension across the mechanically relevant thickness of the membrane is T_m, and only a fraction f reaches the channels, then the slope of the Boltzmann curve is shallower by f. When comparing the properties of MSCs in complex (i.e., biological) membranes it might be useful to consider the dynamic range in the sense discussed above and illustrated in Figs. 5 and 6. Given our ignorance of the constitutive properties of the membrane, the ratio of the highest probability to the lowest might be a better standard than the slope sensitivity for comparing intrinsic channel properties in different systems.

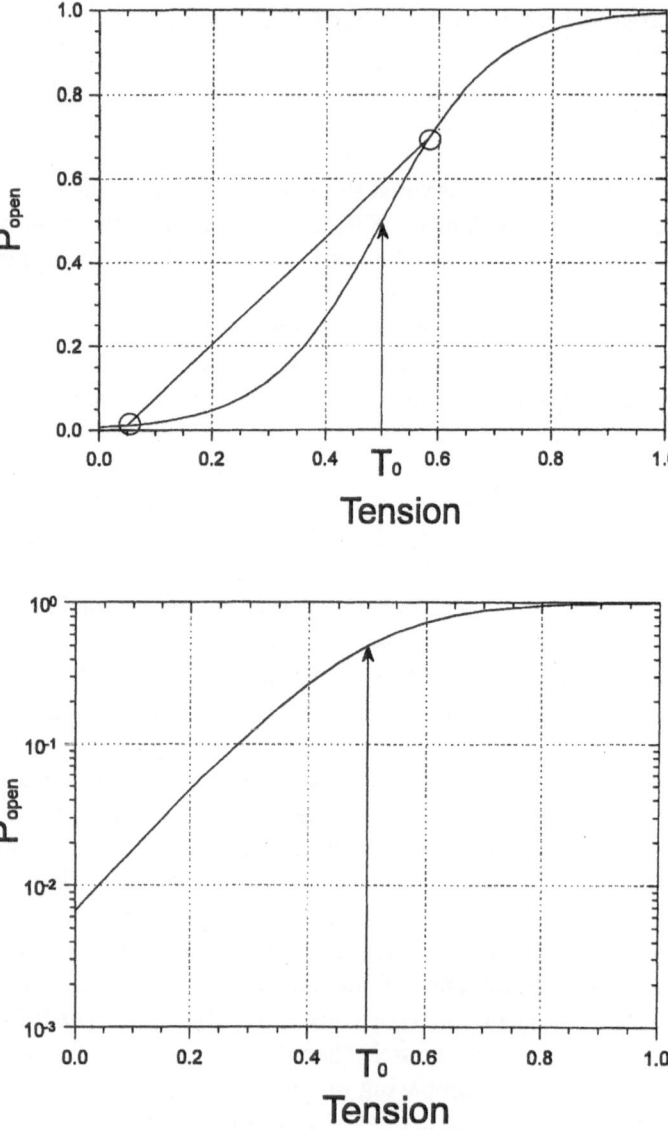

Fig. 5. The probability of being open vs. tension for a two state channel. The Boltzmann relationship is calculated for $T_0=0.5$ and $\Delta A/k_BT=10$ (equivalent to 0.1 dyne/cm/e-fold change in Po, similar to MscL, see below). *Left,* plotted on a linear scale. To place a lower limit on the sensitivity, it is adequate to record activity at only two tensions. *Straight line,* the slope that would be calculated for two observations (*circles*). *Right,* the same P_{open} curve on a log scale, illustrating that for P_{open} below 0.4, a log plot of the data gives a highly linear approximation to the sensitivity

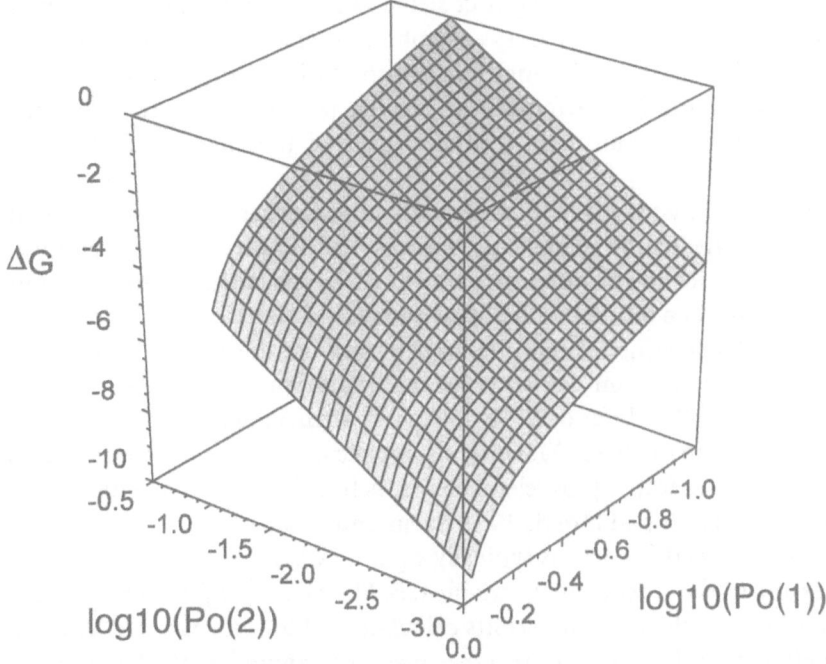

Fig. 6. The free energy difference between the open states of a channel under stimulus conditions 1 and 2 as a function of the probability of being open at two stimulus conditions. $Log_{10}P_0$ values are shown on the x and y axes. ΔG is in units of k_BT

2.1.5
Physical Models for MSC Gating

Returning to the issue of the physical modeling of MSCs, we can ask what is the physical consequence of doing, for example, 7 k_BT of work on a channel. Assume a cylindrical channel of 5 nm diameter and an applied tension of 1 pN/nm (1 dyne/cm). If the channel radius increases upon opening by an amount Δr, then the work done is *tension×change in area*=T×2 πr Δr. Thus, Δr=0.26 nm. The expansion of a 5-nm diameter channel to 5.52 nm involves 7 k_BT of energy with 1 dyne/cm applied. This is actually a small change in dimensions relative to the measurements on molecular sensors and motors. For the transduction channel of cochlear hair cells, the linear change in the channel's dimensions has been estimated to be 2–4 nm (Howard and Hudspeth 1988), and for molecular motors, the conformational power strokes have been estimated to be in the same range (Nishizaka et al. 1995; Svoboda et al. 1995). The above calculation was based on the assumption that the gating transition involved the whole molecule (as cartooned in Fig. 1), whereas it is probably more reasonable to assume that a smaller part of the channel made larger movements.

There are many possible models for how mechanosensitive ion channels are activated by membrane tension. The simplest is that mentioned above, in which the channel appears as a disk embedded in a homogenous lipid membrane for which the open and closed states have different areas (Fig. 4). Such a model may be appropriate for the reconstituted MscL channel cloned from *Escherichia coli* (Sukharev et al. 1994a) or for alamethicin (Opsahl and Webb 1994a) in artificial lipid bilayers. Similar models can be made for MSCs in eucaryotic cells with forces transmitted to the channel from cytoskeletal elements lying parallel to the membrane (see Fig. 2; Lecar and Morris 1993; Sachs and Lecar 1991).

The MSC gating curve is typically sigmoid when plotted against pipette pressure as the independent variable (see Fig. 5), and, as pointed out above, knowledge of the tension permits calculation of the effective ΔA. Unfortunately T is rarely known for two reasons: (a) T cannot be calculated from P, the trans-patch pressure, without knowing the radius of curvature, and (b) in cells the compliance of the various elements that are in series and parallel with the channels is not known. Moreover, neither radius of curvature nor compliance are fixed parameters; both are expected to vary with applied pressure. We should assume that only a fraction of the apparent tension arrives at the channel. Published values of the slope of the activation curves are widely spread and are more likely to represent different cytoskeletal structures surrounding the channel and differences in patch geometry rather than differences in the channels themselves.

In the case of a channel in a lipid membrane, the analysis of forces is simpler than in cells since there is only one phase surrounding the channel (although different stresses may exist in each monolayer as discussed below; Evans and Yeung 1994). For a bilayer in a pipette, the tension, T, can then be estimated from the pressure gradient, P, and knowledge of the radius of curvature using the Laplace equation: $T=Pr/2$, where r is the radius of curvature and T is the tension. The only published case in which this has been done directly in a patch is in a study of alamethicin, an antibiotic that produces multimeric channels with multistate conductance pores (Opsahl and Webb 1994a). In this case the applied tensions were in the range of 0–8 dyne/cm and the probability of occupancy of each conductance level changed by $\approx$8 dyne/cm/e-fold. This slope sensitivity corresponded to an increase in area of $\approx$1.2 nm^2 for the addition of one monomer. In a more complicated system of spherical protoplasts from yeast, Gustin et al. (1988) measured mean currents as a function of pressure. Calculating from their more recent data (Gustin 1991), a vesicle of 0.69 pF subjected to pressures of 0–8 cmHg had a slope sensitivity of $\approx$0.05 dyne/cm/e-fold with a midpoint at 0.7 dyne/cm. This is equivalent to a change in area (closed to open) of about 4.9 nm^2, corresponding to a change in diameter 0.15 nm if the channel were 5 nm in diameter. Sokabe et al. (1991) measured the P_o vs. tension slope sensitivity in patches of chick skeletal muscle and found 3.3 dyne/cm/e-fold. The difference between the very high slope sensitivity of the yeast data compared to the chick muscle probably reflects shunting of tension around the channels in the muscle.

2.1.6
Bilayers as Two Monolayers

The previous discussion of gating models has considered the case of the lipid membrane being isotropic. However, there exists data suggesting that the membrane may need to be considered as two monolayers (Evans and Yeung 1994). Martinac et al. (1990) applied surface active amphipaths with the intent of modifying the lipid structure and the gating of SACs in *E. coli*. The experiments suggested that regardless of which side of the bilayer the amphipaths inserted, SACs became activated. When the amphipaths were inserted into both sides, SAC activity returned to the control levels. Similar data have been reported for MSCs in chick skeletal muscle (Sokabe et al. 1993a).

These surprising effects were modeled by considering the bilayer as two monolayers that could have different tension, with the sum equal to the "far field" tension (that tension created by suction in the pipette; Markin and Martinac 1991). Inserting amphipath molecules into the outer monolayer caused it to expand. If the outer and inner monolayers were limited in the number of available lipids, then expansion of the outer layer would stretch the inner monolayer and thereby activate the channel. An analogy for this coupled expansion effect would be pulling on a pair of elastic ropes (monolayers), and then adding extensions (amphipaths) to one of the them. The load would be transferred to the shorter of the ropes, increasing its tension. When both monolayers contained equal amounts of amphipath, the membrane area increased equally in both halves with no increase in tension in either monolayer and the channel was not activated (Fig. 7).

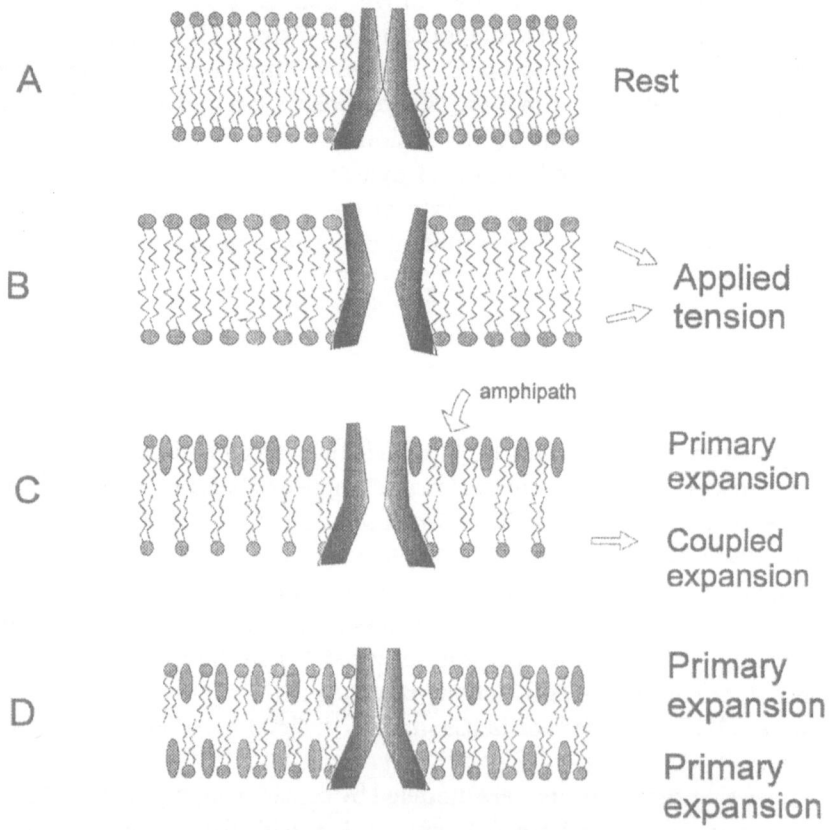

Fig. 7A–D. MSCs in bilayers under the influence of applied tension and amphipaths. **A** The membrane and the channel at rest. **B** The membrane is under tension (induced by patch pressure) that pulls lipid molecules apart and promotes channel opening. **C** An amphipath that prefers the upper monolayer is added. This has the primary effect of expanding the upper monolayer. However, because the number of lipid molecules in each layer is assumed fixed, expansion of the upper monolayer expands the lower monolayer putting it under increased tension. The upper monolayer is not stressed because the amphipaths do the work of keeping the lipids apart. Tension between lipid molecules in the lower monolayer pulls on the channel activating the gate. **D** If amphipaths are introduced to both monolayers, neither experiences a tension increase and channels remain closed

This bilayer expansion theory is based on a number of unrealistic assumptions: (a) that the membrane is constrained from relaxing to its natural curvature following addition of amphiphilic spacers to one monolayer, (b) that the area of the inner and the outer monolayer are equal, i.e., the number of phospholipids in the outer and inner monolayers is unchanged by either the addition of amphipaths or the application of tension, and (c) that the SACs activate with tension in either monolayer.

The channel that Martinac et al. (1990) studied required significant far field tension to be activated even in the presence of amphipaths (T_0 was large; see above). This meant that both monolayers were under far-field tension regardless of the presence of the amphipaths. Similar experiments need to be done on other MSCs with lower T_0 values. If MSCs were found that responded to amphipaths without far-field tension, the theory would have to be reevaluated, for in that case the channel would be under compression in the monolayer containing amphipaths and under tension in the monolayer without amphipaths. If similar results were obtained, it would signify that the channel only responded to tension and not compression! This asymmetry of response would seem to violate the fluctuation-dissipation theorem that states that thermal fluctuations produce the same relaxations as small amplitude driving forces, such as alternating compression and tension.

Markin and Martinac (Markin and Martinac 1991) pointed out some conflicts between the data and the theory. For example, neutral amphipaths produced similar effects, although one might argue that because of intrinsic differences in the inner and outer monolayers, the neutral amphipaths were still asymmetrically distributed. Kubalski et al. (1993) found that the sensitivity to amphipaths did depend upon the lipid environment. Elimination of the naturally occurring major lipoprotein reduced the channel's native sensitivity and its ability to respond to small amphipaths such as chloropromazine and trinitrophenol. Also, they pointed out that the time response of the amphipath effect was hard to explain. Activation by amphipaths was slow ($t_{1/2} \approx 20$ min), they argued, because the amphipaths that concentrated on the *trans* side of the application had to undergo slow diffusion to the opposite side. However, the reversal of *trans* applied effects by *cis* applied amphipaths should be much faster and it was not. Furthermore, addition of the opposite polarity amphipath could initially increase, rather than decrease, the effect.

An alternative explanation of their data, and the one originally suggested, is that amphipaths altered the local membrane curvature and that gating of the MSCs was sensitive to local curvature (Martinac et al. 1990). It is known that amphipathic compounds can alter the intrinsic curvature of lipid bilayers and red cells. However, symmetry of the activation of MSCs by amphipaths on different sides of the membrane would still be hard to explain. The MSC would have to be activated by curvature of either sign. Regardless of the proper theory, the observations of Martinac et al. (1990) are interesting, and the theory (Markin and Martinac 1991) would seem to incorporate the basic elements required to explain the data, even if the assumptions are not satisfied precisely. Their experiments need to be repeated on other systems and measurements made to test the assumptions regarding partitioning and curvature. It will be especially interesting to know if the observations hold true for the MscL channel that has now been cloned from *E. coli*, because that would open the possibility of molecular analysis of the phenomenon. A useful amphipathic tool may be chlorotetracycline. In red cells, the drug can reversibly partition into either the outer or the inner membrane under the control of extracellular Ca^{2+} levels (Riquelme et al. 1982).

The quantitative treatment of bilayers as two dynamic monolayers has recently been published (Evans and Yeung 1994; Yeung 1994) and may have relevance to the dynamic properties of MSCs. When patching bilayers, only the outer monolayer is in contact with the pipette. Tension produced by a sudden change in pressure stretches the outer layer, and because of viscous coupling between the outer and inner monolayers, the inner one is also stretched. However, if there is excess lipid, the fluid inner monolayer will relax with time, leaving all the tension borne by the outer monolayer. A cartoon of this is shown in Fig. 8. The viscous relaxation of bilayer stresses may account for the "adaptation" or inactivation seen with the MscL channel in lipid bilayers (Hase et al. 1995). According to Yeung (1994), the time required for relaxation of the excess density is on the order of, $\tau = (2r_v^2 \ln(2r_v/r_p))/ D$ where r_v is the diameter of the vesicle attached to the pipette (the reservoir of excess lipid), r_p is the radius of the pipette and D is an equivalent diffusion constant that has a value of $\approx 10^{-5}$ cm^2/s. For $r_v = 10$ μm and $r_p = 1$μm, $\tau \approx 0.6$ s, not far from the values observed for MscL (Hase et al. 1995).

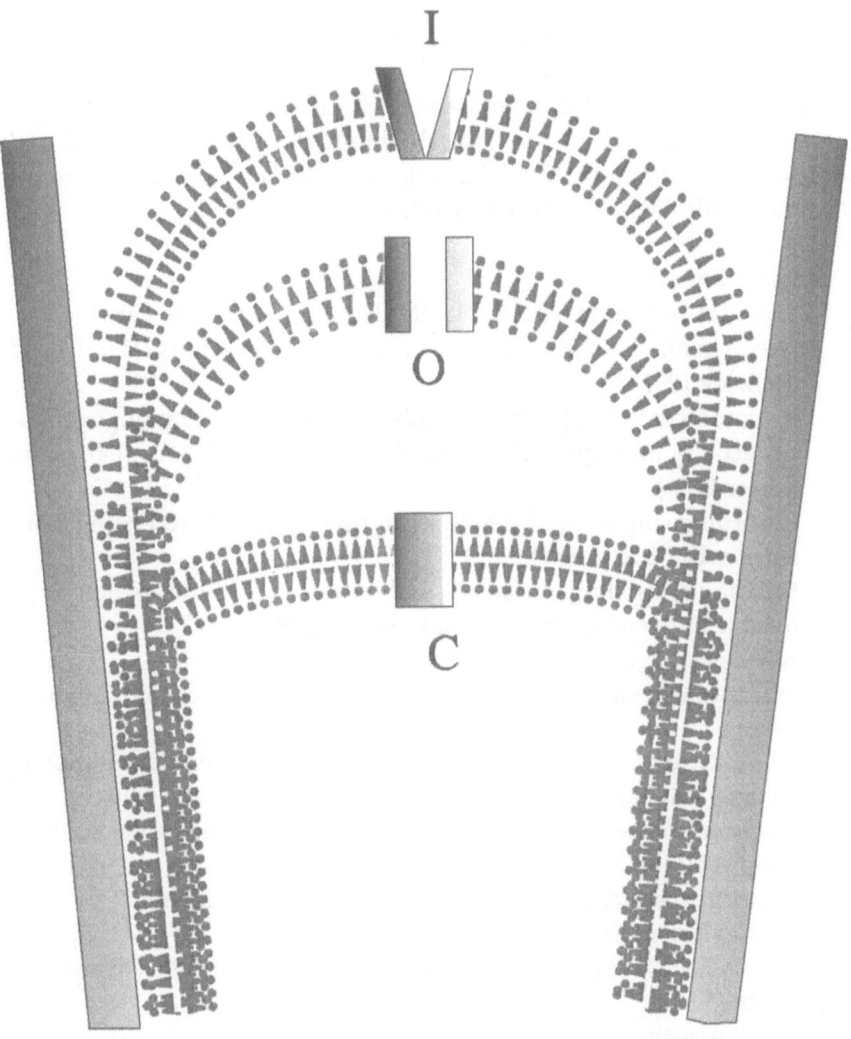

Fig. 8. The effect of monolayer slippage on bilayers and MSC properties. At rest the membrane is not under much tension, is relatively dense, and the MSC is closed (*C*). Following a step of suction, the membrane is pulled upward and stretched against the attachment to the walls of the pipette. Since the outer and inner monolayers are coupled by viscosity, both are stretched causing the MSC to open (*O*). With time, the inner monolayer flows back to normal density relieving its tension and the MSC inactivates (*I*)

3
Methods of Mechanical Stimulation

Given the empirical definition of what constitutes a MSC it is important to understand the forms of mechanical stimulation that can be used and some of their limitations. The following methods have been developed for studying single channels and whole cells:

- Single channels
    - Pressure applied to a patch pipette (Guharay and Sachs 1984)
    - Use of gentle patches to reduce cytoskeletal alterations (Hamill and McBride 1992)
    - Hypotonic stress applied to a whole cell while recording single-channel activity through a cell-attached patch (Ubl et al. 1988; Falke and Misler 1988)
    - Recording single-channel activity during whole-cell recording from small cells (Gustin et al. 1988)
    - Inflating or swelling nucleated patches (these are like extremely large outside out patches; Paoletti and Ascher 1994)
- Whole-cell currents
    - Direct mechanical strain (Davis et al. 1992; Wellner and Isenberg 1994)
    - Cell inflation or deflation through a broken-patch whole-cell pipette (Morris and Horn 1991b; Wan et al. 1995; Gustin et al. 1988)
    - Anisotonic stress (see Oliet and Bourque 1993; Ackerman et al. 1994; Popp et al. 1992; Filipovic and Sackin 1992)
- Reconstituted systems: bilayer, liposomes, membrane blebs
    - E. coli MscL in liposomes (Hase et al. 1995)
    - Bovine ENaC (epithelial Na^+ channel) in planar bilayers (Awayda et al. 1995)
    - E. coli in blebs (Saimi et al. 1993)

3.1
Single Channels

3.1.1
Pressure Stimuli

Single channel events elicited by pipette pressure are readily identifiable as MSCs when the response is rapid, which generally it is. The key variable is tension and not pressure. While tension is rarely measured, the pipette pressure is generally not the same as the trans-patch pressure and the forces and pressures within the cytoskeleton of the patch are unknown. It is not universally recognized that even the reported pipette pressures are commonly in error since there may be offsets due to residual capillary pressure, fluid column height or errors in the pressure source (e.g., Morris et al. 1989). The best way to establish a zero pressure reference is to adjust the offset of the pressure source until there is no flow through the pipette tip as visualized by the movement of small particles in the bath.

3.1.2
Tension vs. Pressure

A second factor affecting the stimulus magnitude is that the membrane tension is a function of the trans-patch pressure and the radius of curvature. There is currently no way to know the curvature without directly visualizing and measuring the geometry of the stimulated patch (Sokabe et al. 1991). The pipette tip radius is not a good indicator of the relevant radius for two reasons: the patch generally does not seal at the tip but further up (Sokabe and Sachs 1990; see Fig. 3), and the radius of curvature is a function of the trans-patch pressure. The latter effect has been analyzed by Sokabe and coworkers (1993b) assuming that the membrane is a two dimensional elastic sheet. The result of this derivation is that the tension is given implicitly by:

$$P = (4\,K_A/r)(T/K_A)^{3/2}/(1+T/K_A) \tag{3}$$

where P is the pressure, K_A is the area elasticity, T is the tension and r is the radius of the pipette. Some representative relationships are shown in Fig. 9. The tension is a sensitive function of the patch radius as shown in the upper panel. An applied pressure of 10 mmHg can produce tensions of 3.9,

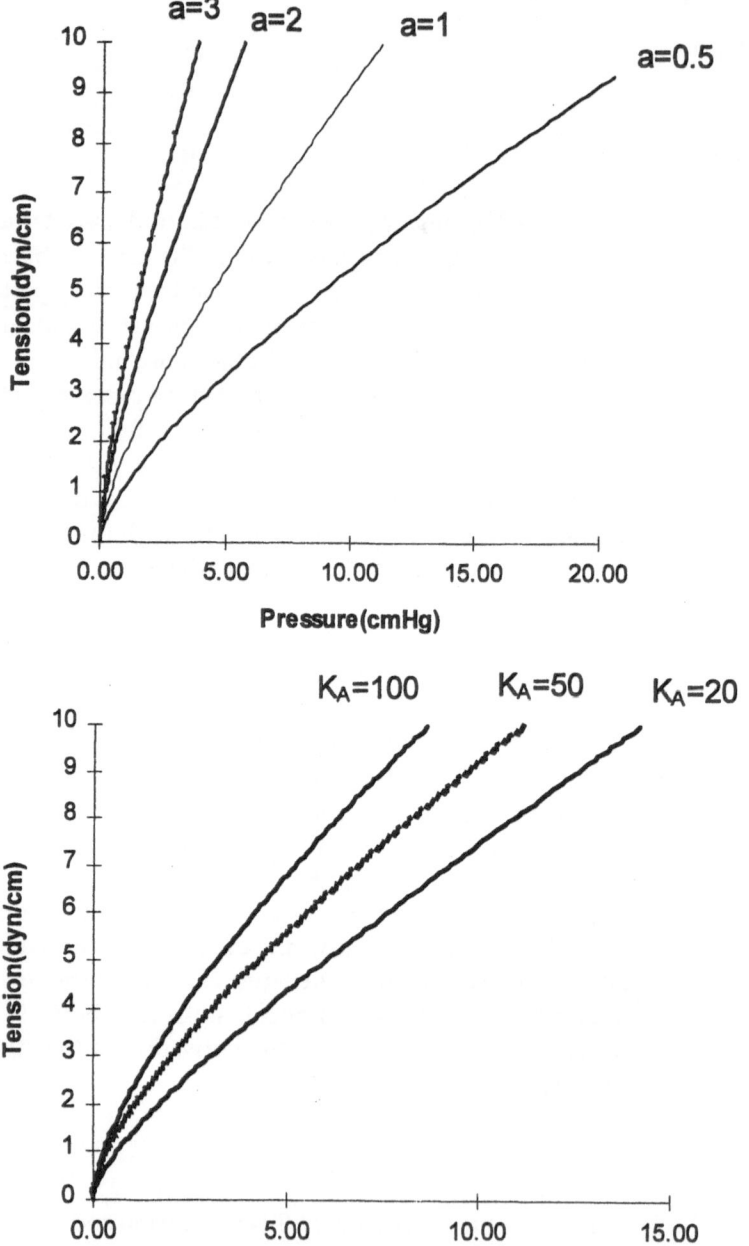

Fig. 9. Response of an elastic patch to pressure. *Above*, the tension for different pipette radii (*a*) with a membrane having an area elastic constant, K_A=50 dyne/cm. *Below*, the tension for a fixed radius of 1 µm and different elasticities

2.9, 1.8 or 1.2 dyne/cm depending upon the patch radius (3, 2, 1, 0.5 μm, K_A=50 dyne/cm). Since outside-out patches in general have smaller radii (Ruknudin et al. 1991; Sakmann and Neher 1983), MSCs in such patches tend to be less sensitive to pressure. Patches of larger dimensions should exhibit MSCs with higher pressure sensitivity. This has important consequences for assessing SAC activity. If a reagent is tested for its effects on MSCs and this reagent happens to alter the mechanical properties of the cytoplasm so that larger patches are formed, SAC sensitivity may appear to have increased when in fact it is unaltered or even decreased. In such a case, the area elasticity would probably also change, making things even more uncertain.

In the elastic sheet model of a patch described above, the tension depends on the pressure in a nonlinear manner. Allowance for a resting tension was left out of the calculation above, although that can be included (Sokabe et al. 1993b). The existence of resting tension can be shown by observing the flat (rather than wrinkled) profile of a patch at zero pressure (Sokabe and Sachs 1990) and by extrapolation of a plot of dA/dT to zero tension (Sokabe et al. 1993b). Resting tension may be brought about by adhesion of the membrane to the wall of the pipette (estimated to be 0.5–4 dyne/cm for lipids; Opsahl and Webb 1994b) or by stresses normal to the membrane arising from a distorted cytoskeleton (see Fig. 3 and Sokabe and Sachs 1990a). Measurements of patch mechanics reveal an additional complexity – the bilayer flows under stress so that the amount of bilayer within the patch is not constant (Sokabe et al. 1991).

3.1.3
Sharing of Tension

A less easily resolved issue in defining the stimulus is that the membrane is a complex structure composed of a lipid bilayer, transmembrane proteins, extracellular matrices, and parallel and series membrane bound cytoskeleton coupled to the interior cortical cytoskeleton through time and force dependent links. The proportion of the applied stress that reaches the channel can be time, velocity and history dependent (Sokabe et al. 1993c; Hamill et al. 1992). A consequence of this lack of controlled stimulation is that any intervention that appears to affect channel activity may have its origin in modulation of the cytoskeleton or extracellular matrix (Chen et al. 1994). Any intervention that affects the way in which force reaches the channel or affects the size of the patch and its curvature can affect estimates of both the channel sensitivity in terms of pressure (or mean tension) as well as estimates of the channel density.

As has been discussed above, some properties of MSCs are time dependent. Proper studies of such phenomena require fast response pressure sources. Steps of pressure can be created by using solenoid valves to switch between fixed pressure reservoirs or by employing feedback systems (Sachs 1987; Sokabe et al. 1991; McBride and Hamill 1992; Small and Morris 1994). These may be oil or air filled systems and they can have risetimes as short as a 1 ms. Despite the tight control one can have over the pipette pressure, one must be cautious about the trans-patch pressure. Although the atmospheric pressure of the bath is fixed, the pressure drop and mechanical stresses created in the cytoplasmic portion of the patch are not clear (see Fig. 3). The cytoplasm is viscous and resists deformation. A step of pressure can generally be guaranteed *not* to produce a step in tension. Furthermore, one needs to ask, "In what component of the cortex does the tension change in a prescribed way with a prescribed stimulus?" Small and Morris (1994) showed that SICs and SACs in the same patch may have very different response latencies. This result suggests coupling to different cytoskeletal components, but it may also reflect different inherent sensitivities (Morris and Horn 1991b). Working with MSCs is much more difficult than working with voltage or ligand gated channels because there is no simple stimulator.

3.1.4
Hypotonic Stress Observed Through Single-Channel Activity

Hypotonic swelling has been used as an alternative or a supplement to stimulation by suction (Falke and Misler 1989; Filipovic and Sackin 1992; Sackin 1989). The bath can be diluted (or an inert osmolyte removed) to swell the whole cell, and in cell-attached mode MSC activity is observed. By applying suction to the pipette and observing whether the channel activity increases one can test whether the observed activity arose from MSCs, as opposed to some other volume-activated ion channel. Without applying suction, it is unclear whether one is observing MSC activity or some other channel (Grunder et al. 1992). Since it is known that volume stress can trigger second messenger cascades, and since some MSCs can be activated by second messengers (Vandorpe et al. 1994; Rothstein and Mack 1992; Widdicombe et al. 1991; Kim et al. 1995), some uncertainty remains as to the contribution of different mechanisms to the observed activation.

3.1.5
Whole-Cell Stretch

Unfortunately, it is not possible to reliably record single-channel activity evoked by whole-cell stretch in the cell-attached mode. Sealing of the membrane to the glass pipette severely alters the distribution of stress and it is not uncommon for the patch and its adherent cortex to physically separate from the rest of the cell while maintaining a seal as shown in Fig. 3. Thus, in general it would not be possible to apply direct mechanical stress to cells (such as pulling on the ends) while observing the activity of MSCs in a cell-attached patch. However, it is sometimes possible to observe single-channel activity (at low resolution because of the low source impedance) when recording currents from small whole cells or vesicles (Gustin et al. 1988).

3.2
Whole-Cell Currents

3.2.1
Direct Mechanical Strain

In principle, the simplest form of mechanical stimulation should be stretching a cell. Unfortunately this is difficult because, in most preparations, it is difficult to attach pulling probes to the cell without producing local stress or even damage. Skeletal muscle is one preparation that comes with attachments in the form of tendons that should make these experiments possible. Unfortunately for the field, although stretch-induced increases in cell Ca^{2+} have been observed (Snowdowne 1986), these do not depend on extracellular Ca^{2+} suggesting that sarcolemmal channels are not involved (MSCs in intracellular organelles have not been ruled out, however). Smooth muscle cells from blood vessels and the urinary bladder have been stretched (Davis et al. 1992; Wellner and Isenberg 1994, 1995) by holding them with suction pipettes or wrapping them around glass probes and voltage clamping with a third pipette. In these cases stretch induced inward currents have been recorded that appear to match the expected properties seen with single channels in cell-attached recordings.

Cardiac physiologists have spent a lot of time exploring ways to attach probes to cells to measure the length dependence of various properties (Tung and Parikh 1993; Tung and Zou 1995; Brady 1991; Gannier et al.

1994; Wellner and Isenberg 1994). The best methods to date use carbon fibers (White et al. 1993) or sticky glass probes (Sasaki et al. 1992; Tung and Parikh 1993; Palmer et al. 1996) that adhere to the sarcolemma. The work with carbon fibers (Gannier et al. 1993) suggests that, for reasons not understood, the strain is rather uniform – when cells are stretched, the sarcomere spacing doesn't vary significantly below the attachment site of the probe. Presumably this means that there is significant rigidity of the contractile apparatus across the fiber diameter.

There have been a few papers on the effects of whole-cell stretch under voltage clamp conditions (Sasaki et al. 1992; Wellner and Isenberg 1994; Davis et al. 1992). In these reports, there was a stretch-induced cationic current that reversed at $\approx$–15 mV. In a different approach, whole-cell mechanosensitive currents have been evoked by pressing on a spherical cell with the side of one pipette while voltage clamping with another (Hu and Sachs 1994, 1995, 1996).

Although voltage clamping is preferred to obtain quantitative data from excitable cells, current clamp experiments and the use of Ca^{2+} indicators have shown stretch-induced depolarization and increases in Ca^{2+} in guinea pig heart cells (White et al. 1993). Similar results were obtained in chick heart cells (Sigurdson et al. 1992), endothelia (Diamond et al. 1994; Sigurdson et al. 1993), glia (Charles et al. 1991), osteoclasts (Xia and Ferrier 1995), nodose ganglia (Sharma et al. 1995) and epithelial cells (Boitano et al. 1994). For example, in the case of the heart cells (Sigurdson et al. 1992), mechanical deformation induced Ca^{2+} waves initiating from the site of stimulation. These effects were dependent upon extracellular Ca^{2+}, blocked by Gd^{3+} and by culturing cells under conditions that suppressed SAC activity. That data also suggested the SACs feed Ca^{2+} to a restricted volume between the sarcolemma and the sarcoplasmic reticulum so that the relatively small Ca^{2+} flux through SACs (Yang and Sachs 1990) was rendered capable of initiating Ca^{2+} release.

In a different version of this Ca^{2+} experiment, endothelial cells (Naruse and Sokabe 1993) and lung cells (Wirtz and Dobbs 1990) have been exposed to shear stress induced by fluid flow or by growing them on rubber sheets subjected to radial or linear stresses. In the case of the lung cells, however, the Ca^{2+} elevation did not appear to be caused by plasmalemmal MSCs because the response was not sensitive to extracellular Ca^{2+}. In contrast, by pulling on fibroblasts with a sticky pipette, Hagmann and coworkers (Hagmann et al. 1992) were able to cause endosomal release in way that depended upon external Ca^{2+}. This variety of ways in which mechanical stress can produce cellular responses points out the need to do direct patch clamp experiments to verify MSC involvement, at least until such time that specific pharmacological tools become available.

3.2.2
Inflation

Hydrostatic pressure is sometimes used as a stimulus in broken-patch whole-cell clamp experiments by inflating the cell through the clamping pipette (Langton 1993; Morris and Horn 1991b; Wan et al. 1995; Doroshenko and Neher 1992). It seems possible to inflate the cell this way, but the distribution of stress is very unclear, as may be envisaged in a model experiment of injecting water into a ball of gelatin. Complications may also arise from dilution of the intracellular compartment, although that seems no worse than the dilution that normally accompanies broken patch dialysis. A possible artifact arises from the flow of solution through the pipette that can alter tip potentials via electro-osmotic effects. Langton (1993) reported that pipette tips often seemed to become mechanically blocked even when there was no concomitant rise in electrical series resistance.

3.2.3
Hypotonic Stress

Since hypotonic swelling would appear to be a gentle way to deform cells, it has been the choice of many investigators. Unfortunately, swelling may have a host of effects on different systems. It leads to dilution of intracellular components, volume stress on the cytoskeleton, stress on organelles, possible increases in plasma membrane area (Wan et al. 1995) and direct activation of some channels including MSCs. We will not review the vast literature on volume regulation, but in summary, the most common electrical effect of hypotonic stress is an activation of Cl^- pathways (Grunder et al. 1992; Ackerman et al. 1994). The difficulty of interpretation is pointed out in systems such as *Xenopus* oocytes that posses cationic SACs, but exhibit anionic volume activated currents and no evidence of swelling activated SAC currents (Ackerman et al. 1994). In chick heart cells, strain-induced whole-cell currents were cationic but volume-stress-induced currents were anionic (Hu and Sachs 1996). MSCs may be a trigger signal for volume regulation (Chen et al. 1996), but their net current is overwhelmed by the amplified anionic component.

3.3
Reconstituted Systems:
Bilayer, Liposomes, Membrane Blebs

In principle, reconstitution into lipid membranes ought to be a good way to test isolated MSCs. This has worked well for the MscL channel from *E. coli* (Hase et al. 1995; Sukharev et al. 1994a). As discussed Sect. 2, a bilayer may be subject to adaptation because of shear stresses in the bilayer, but a more serious conceptual limitation is that there is no evidence that the bilayers are under any tension in eucaryotic cells. Dai and Sheetz's (1995) measurements of neuronal growth cone membrane tension suggest that in-plane tension in the bilayer of neurons is too weak to have a significant effect on MSC channel gating. Thus, if one inserted a prototype MSC into a bilayer (planar or vesicular) and showed activation with stress, the most likely conclusion is the presence of an artifact: the channel doesn't see lipid tension in vivo. This difficulty is serious when attempting to evaluate a presumptive cloned MSC.

Equally serious problems exist when using planar lipid bilayers instead of lipid vesicles. Because of the wetting of the support, planar bilayers are under a substantial resting tension in the range of 1–5 dyne/cm, depending upon composition (Elliott et al. 1983; Gruen and Wolfe 1982; Ring and Sandblom 1988; Ring 1992). Thus, a MSC in a planar bilayer would tend to be activated at "rest." It may be significant that the resting P_{open} of epithelial Na^+ channels in planar bilayers was near 0.5 (Awayda et al. 1995). A second problem with using planar bilayers for MSC studies is that they are basically tension clamped. Because of the excess lipid available in the torus (material around the supporting structure, in more exact terms, the Plateau-Gibbs border), a change in the transmembrane hydrostatic pressure gradient will not change the membrane tension significantly, but will draw in more material having the same density (Fig. 10).

Pressure gradients will bow the membrane, but there is almost no membrane stretching. Laplace's law is satisfied with changes in radius so as to maintain constant tension, $r=2T/P$. Thus, it is unclear how to explain the results of Awayda et al. (Awayda et al. 1995) in which they showed a small increase in P_{open}, with a transmembrane pressure gradient. Perhaps the channels were sensitive to membrane curvature. In vivo observations

Fig. 10. Cartoon of the Plateau-Gibbs border (torus) of a planar lipid bilayer. The thickness of the partition between chambers is actually much wider than indicated in the drawing. The empty spaces between the partition and the bilayer are filled with micelles, and possibly the lipid solvent, that serve as a buffer of excess material to flow under the influence of tension caused by bowing. The contact angles of the border/bilayer provide resting tension that stretches the bilayer flat. As the membrane bows under a hydrostatic gradient, the contact angles on the right and left sides change slightly providing for a second order increase in tension

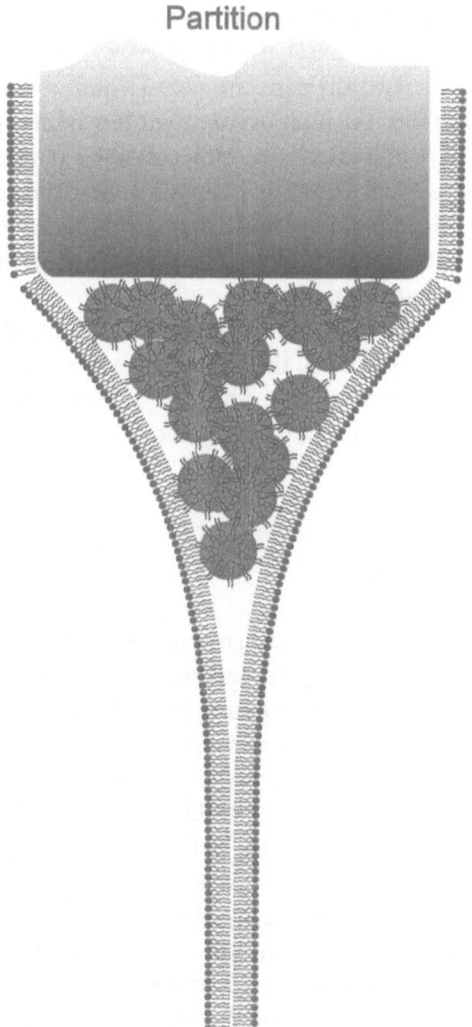

Partition

from the collecting tubule show that in some patches the epithelial Na$^+$ channel gating is stretch sensitive (Palmer and Frindt 1996).

Finally, there is a general problem in applying tension to bilayers: the partitioning of materials into the membrane increases as the density decreases (Gruen and Wolfe 1982). This may account for the increase in peak currents observed to follow repeated stress of membranes containing

MscL channels (Hase et al. 1995). With each application of tension, more monomers enter the membrane to form channels that are then available for activation with the next stimulus. Any lipophilic molecule will partition into the membrane, including contaminants, and the magnitude of the effect increases with molecular size (Gruen and Wolfe 1982; Lehtonen and Kinnunen 1995). It is possible that tension induced partitioning of fatty acids (Kim 1992) and other amphiphiles may affect the gating of MSCs and perhaps make channels that are not intrinsically MSCs appear to be mechanosensitive.

In summary, the lack of a simple stimulator for MSCs has contributed to slow progress in the field. Awareness of the potential errors and limitations of the methodology, however, should improve the quality of experiments and permit further progress.

4
Structure

What structural information is available about MSCs? The answer is: very little. Nevertheless, several of the rather disparate channels that exhibit mechanosensitive gating at the single-channel level in some preparations have been cloned, and these allow us to assert that Nature's requirements for MSCs must be rather catholic. The cloned channels in question are: MscL from bacteria (Sukharev et al. 1994a), the NMDA channel, a glutamate activated nonselective cation channel (Paoletti and Ascher 1994), smooth muscle Ca-activated K channels (Dopico et al. 1994; McCobb et al. 1995), a G-protein regulated K-permeant inward rectifier, GIRK (Krapivinsky et al. 1995; Pleusamran and Kim 1995). Of these, only the MscL was cloned because of its mechanosensitivity. The small peptide subunit (≈15 kDa) coded by the MscL gene forms multimers with a central pore region, but the putative hexameric structure (Sukharev et al. 1996a) differs from either the five-subunit heteromultimer that constitutes the NMDA channel, the tetramers (presumably) of the Ca-activated K channels (McCobb et al. 1995) and the four-subunit heteromers of GIRK (Krapivinsky et al. 1995). The MscL peptide itself appears not to be homologous to any other peptide although its 3 nS conductance is similar to that of the bacterial porins. At this stage it seems fair to say that there are no stringent requirements on global structure for a channel to exhibit some mechanosensitivity. Since it is unknown whether any of these channels uses its mechanosensitivity in situ, it is possible that their diverse structures are irrelevant for mechanosensitive gating. Weak mechanosensitive gating may occur in the same manner as weak voltage dependent gating: simply by spanning membranes with intense electric fields, channels become susceptible to membrane potential effects. Weak voltage sensitivity is physiologically trivial and requires no differentiated protein structures, such as the recurrently charged S4 segments of "true" voltage-gated channels. Likewise, it may be that the regular gating motions of weakly MSCs cannot be entirely shielded from mechanical interference, so that intense mechanical stimuli can affect these gating motions.

4.1
Mechanosensitivity of Ligand-Gated Channels

Extracellular ligands and cytoplasmic or membrane-delimited second messenger molecules are all, in the final analysis, ligands. Various channels can be activated via ligands of second messenger systems as well as by membrane stretch. Such channels include the K-selective *Aplysia* S channel (Vandorpe and Morris 1992), and cation channels in osteoblasts (Duncan et al. 1992), hepatocytes (Bear and Li 1991; Bear 1990) and kidney cells (Marunaka et al. 1994; Verrey et al. 1995). Intracellular calcium qualifies as both ligand and second messenger. Ca-activated K channels in smooth muscle (Dopico et al. 1994) and kidney (Taniguchi and Guggio 1989) exhibit mechanosensitivity. There remain questions, however, as to whether the CaK channels themselves are stretch sensitive or whether they are coupled to "true" stretch activated channels through a direct Ca^{2+} flux or Ca^{2+} flux mediated via Na/Ca exchange. Na Glutamate is an extracellular ligand; and one class of glutamate-activated channels exhibits mechanosensitivity (Paoletti and Ascher 1994). It becomes a question of definition as to whether one should call these mechanosensitive channels that are also ligand sensitive, or ligand sensitive channels that are also mechanosensitive. Where knowledge permits, the name should first reflect the most physiologically relevant source of free energy, and should be modified by the alternate source, but our understanding of the multiple roles of channels is still too rudimentary for us to follow this course with any accuracy.

One example about which we do have some information is the NMDA channel. This channel is known to respond to glutamate in a physiological context, and it cannot be opened by stretch alone (Paoletti and Ascher

1994). Thus, we would call it a ligand gated channel that is modulated by tension. This is straightforward because ligand-binding seems to be more effective than increased membrane tension in lowering the free energy difference between open and closed states. But what of the *Aplysia* S-channel (Vandorpe et al. 1994; Vandorpe and Morris 1992)? It is physiologically activated by arachidonic acid metabolites and inhibited by an A-kinase; the levels of these "ligands" are controlled via the neurotransmitters FMRFamide and serotonin, respectively. Under patch clamp conditions, however, it can be activated by stretch alone, and the maximal effect of stretch greatly exceeds that of the arachidonic acid metabolites. SA K channels akin to the S-channel are ubiquitous in molluscan neurons (Bedard and Morris 1992; Morris 1992), but it is only in identified neurons of *Aplysia* that we know what neurotransmitter systems control the channels. The channels do not readily activate with mechanical stimuli in situ (Morris and Horn 1991b); yet, "SA K channel" is currently the best designation we have for these channels. We may currently be designating other channels as "MS channels," perhaps even the *Xenopus* oocyte MS channel (Steffensen et al. 1991) or the bacterial MscL, simply because the primary physiological stimulus has not been identified.

4.2
Role of Auxiliary Structures

The most distinctive features of specialized mechanoreceptor cells, such as the hair cells of the cochlea (Ashmore 1991; Hackney and Furness 1995), are not the channel properties, but the elaborate mechanical impedance matching structures that are used to apply force to the MSCs. It would appear that the extraordinary sensitivity of the hair cells, limited only by thermal noise (Denk et al. 1989; Denk and Webb 1992), arises from exquisitely orchestrated focusing of mechanical energy on channels that, in and of themselves, are not much different from those that occur in nonspecialized cells (Sachs 1988). The mechanical impedance matching devices of hair cells are extracellular and are dependent on highly specialized cell structure, and the nonspecialized cells do not appear to possess such specializations. It is reasonable to ask, then, whether MSCs require linkage to elements of the intracellular membrane skeleton and, conversely, is a linkage to the membrane skeleton sufficient to render a channel mechanosensitive? The answer to the first questions is "no," and to the second question is "probably not."

In nonspecialized cells, where MS channels have been identified by single-channel recording, *macroscopic* cellular structures are clearly *not* required for channel mechanosensitivity. The general membrane/cortex is a favorable environment, although it is not yet clear what the relevant forces are and how they are coupled to external stimuli. Clearly, patch formation stresses are not the relevant physiological input! The specialized mechanoreceptors such as Pacinian corpuscles, Meissner cells, muscle spindles and hair cells show a profound attention to structural specialization (Akoev et al. 1988). Coupling to these structures seems directed at selecting certain stresses and rejecting others. In the case of hair cells, stimulation along the axis of short to long cilia is extremely effective in producing an output, whereas stimulation normal to that axis produces a minimal response (Hudspeth 1989).

4.3
Activation Through the Bilayer

The bacterial channel, MscL, reconstituted into a lipid bilayer, gates in essentially the same way that it does in protoplasts or spheroplasts (Sukharev et al. 1993; Hase et al. 1995). In this case, at least, the need for a connection to the membrane skeleton can be ruled out. It is worth noting, however, that the MscL channels in bilayers appear more sensitive to tension than they are in protoplasts. If one makes the (dangerous) assumption that the patch geometry is the same in both cases, in the more complex membrane structure of the protoplasts some tension skirts the channel. A different channel, the antibiotic alamethicin, which forms multimeric, multistate, channels in liposomes, exhibits mechanosensitive gating at all conductance levels (Opsahl and Webb 1994a), although there is probably no physiological value to possessing this capability.

4.4
Linkage to the Membrane Skeleton

It is known that a variety of well-characterized membrane channels and transporters are tethered to the membrane skeleton either via ankyrin (Lambert and Bennett 1993) or directly (Rotin et al. 1994). Certain voltage-dependent K channels are anchored to members of the post-synaptic density family of proteins (Kim et al. 1995). The fact that these anchored proteins have not been characterized as mechanosensitive does not prove that they are insensitive – the question remains open. Certainly, however, there is no need to invoke mechanosensitivity as a reason for anchorage. Transporters may anchor to the membrane skeleton in order stay fixed at some cellular locale or, alternatively, the transporters may serve as anchors for the membrane skeleton. In erythrocytes, the anion exchanger Band 3 supposedly anchors spectrin close to the membrane for mechanical reinforcement of the cell. A distinct advantage for coupling to the membrane skeleton may occur in transporters which deliver metabolites to, say, skeletal muscle, but it is probably counterproductive for many anchored membrane enzymes to allow their activity to be influenced by mechanical stress. Thus, in asking how and if specialized channel/membrane skeletal linkages might be used by MSCs to convey force to the channel, we might well also ask what features can make such linkage stress-proof. It will probably be necessary to reconstitute such assemblages in order to explore these issues.

4.5
Evidence for Cytoskeletal Involvement

When MSCs were first described in muscle cells, the cells were treated with tubulin and actin reagents to test involvement of the cytoskeleton. It was shown that tubulins had no significant effect, but cytochalasins increased the channels' stretch sensitivity (Guharay and Sachs 1984). Clearly then, neither tubulin nor f-actin could be implicated as necessary for mechanosensitive gating (however, membrane bound actin may not be sensitive to cytochalasin since its turnover rate is slow). It was postulated that MSCs were normally linked into a component of the membrane skeleton parallel to the actin network, so that when actin was depolymerized, stress was transferred to the channel-linked component. The identity of that component has not been resolved, although it would appear not to be dystrophin since MSCs are active in dystrophic muscle cells (Franco et al. 1991; Franco and Lansman 1990; Haws and Lansman 1991; Lansman and Franco 1991; Franco-Obregon and Lansman 1994). Recent data from *C. elegans* suggests that mechanosensory channels incorporating the product of the gene, *mec*-2, may be linked to tubulin (Huang et al. 1995) and to the extracellular matrix structures (Driscoll 1996).

For eucaryotic systems, is it necessary to invoke a series coupling to the cytoskeleton? Or, or can forces be transmitted directly through the bilayer? While there is no a priori reason to reject coupling through the bilayer, there are no data to suggest that the bilayer is under significant stress except during lysis (Sheetz and Dai 1996). Arguing against significant strain in the bilayer is the fact that the elastic constant of patches is much less than that of lipid membranes (Sokabe et al. 1991). Because adhesion of the membrane to the glass of the pipette tip produces a resting patch tension that tends to unwrinkle folds, patched membrane should be more sensitive to lipid stress than an intact cell, and yet the stress seems not be borne primarily by the bilayer. A similar conclusion can be reached by analysis of cell swelling induced by hypotonic stress. With swelling, the increase in *apparent* cell area is much more than can be sustained by the 3%–5% maximal distention of a lipid membrane and, where there is no capacitance change, we must argue that changes in apparent area are primarily brought about by the unfolding of wrinkles produced by the cytoskeleton. Although there are (model-dependent) ways to assess membrane tension in intact cellular membranes (e.g., Dai and Sheetz 1995), it is still not possible to measure bilayer tension under reversible strain in such a way as to determine the residual fraction of cortical stress (i.e., the membrane tension) borne by the bilayer.

It has been repeatedly shown that MSCs can retain their mechanosensitivity in excised patches, and it is known from electron microscopy that excised patches have a cytoskeletal structure (Ruknudin et al. 1991). A suggestive bit of evidence that elements of the membrane skeleton may be required even in excised patches, is that when patching membrane blebs, structures in which much of the cytoskeleton is disrupted (Tank et al. 1982), MSC activity is generally absent, even when intact adjacent membrane does show activity (Sachs 1988). Vesicles created by formaldehyde treatment of *Xenopus* oocytes show little cytoskeletal structure, yet they do retain (modified) MSC activity (Zhang et al. 1996). It will necessary therefore be to carefully document the cytoskeletal protein content of these vesicles.

A different kind of indirect evidence for cytoskeletal involvement was presented by Sokabe et al. (1991), who measured the capacitance, elasticity and channel activity of patches. The data showed that the applied pressure increased the membrane capacitance and that this increase could be accounted for by a flow of lipids into the patch. In these experiments, the location of the patch attachment to the pipette did not change under pressure, and, since the lipids could flow, there seemed little alternative to assuming that the membrane skeleton must account for the long range order that stopped the patch from moving and accounted for the patch's measurable elasticity. When SAC activity was measured as a function of the calculated tension (with the radius of curvature of the patch taken to be the value seen in the light microscope), patches of different curvature had the same sensitivity suggesting that the tension activating the channels was, in fact, proportional to the calculated tension, and that it was this tension that also accounted for the change in area. The elastic constant for the patch was a soft ≈50 dyne/cm, far smaller than the elastic constant of lipids that are ≈500–1500 dyne/cm (Evans and Needham 1987). Thus, the stress that activated the channels seemed to be borne by cytoskeleton. The finding that the elastic constant of skeletal muscle patches was insensitive to cytochalasin (Sokabe et al. 1991) suggests that actin structure was already disrupted by the act of patch formation (Small and Morris 1994) and that other membrane skeletal elements, perhaps spectrin, constitute the elements that transfer mechanical energy to the channel.

These experiments point out potential difficulties in the interpretation of data on patch mechanics because fibrous components may be folded or buckled at one tension and hence invisible to small stretches (you can pull a chain, but you can't push it!). If one were to remove a stress bearing element of the membrane skeleton from a patch at constant pressure, the apparent patch area would increase. However, the elasticity, $K_A=dA/dT$, might be higher or lower than the control, depending upon the constitutive properties of the remaining structures (Fig. 3). For example, one might compare the elasticity of normal and dystrophic muscle to estimate the load borne by dystrophin. If dystrophin was the dominant load bearing element, a patch formed from normal cells would reflect dystrophin's elasticity. A patch from dystrophic cells would form with the load distributed on elements that were not previously loaded. These elements may be more or less compliant than dystrophin (see Fig. 11).

The role of the extracellular matrix in MSC function has been barely touched. Inhibition of growth of the matrix improves seal formation in a renal cell line but does not block MSC activation (Izu and Sachs 1991). Similar results on improved sealing and additionally increased vesicle formation were seen in smooth muscle (Olesen 1995). These data suggest a mechanical involvement of the matrix, but no direct measurements are available.

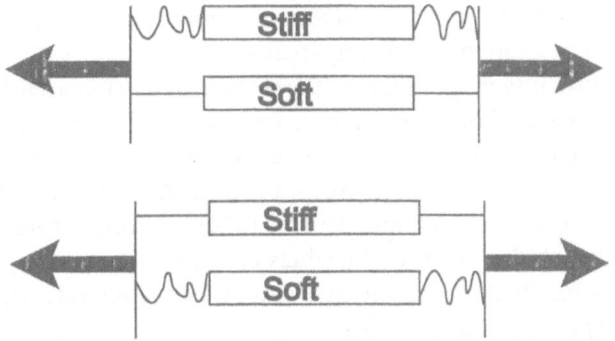

Fig. 11. Two units of membrane subjected to tension (*arrows*) slack in different coupling elements. Removal of the stiff element (*upper panel*) does not affect the elasticity of membrane until the stretch is sufficient to remove the slack of the upper element. This membrane appears soft for small deformations. If a patch is formed from these two example membranes, and the load-bearing element is removed while the patch is under observation, we observe an increase in apparent patch area giving a clue as to the loss of a load bearing-element. However, if a patch is formed from the membrane in which the load bearing element has already been removed, there is no indication that the original load-bearing element has any role to play. Currently we cannot identify these various components with any particular proteins of the cytoskeleton

4.6
Mechanically Fragile Aspects of MSC Channel Behavior

The mechanical properties of patches are labile and history dependent (Small and Morris 1994; Hamill and McBride 1992). Using fibroblasts grown on rubber sheets, Pender and McCulloch (1991) showed that stretch caused f-actin to nearly disappear within 10 s and to reappear at twice control levels when the cells were left to rest for 1 min. Patch formation and stretch stimulation of cells will probably have qualitatively similar influences on f-actin disassembly (Small and Morris 1994; Sokabe et al. 1993c).

Unfortunately, the cytoskeleton cannot usually be removed acutely from a given patch. Inevitably, therefore, patches are sampled from membrane whose mechanical status is dependent on a variable state of the underlying cytoskeleton. Moreover, the size, and hence the radius, of a patch can vary as well as the constitutive mechanical properties; variation in size creates different tensions for the same pressure (Sokabe et al. 1993b), regardless of any local variations in the constitutive properties. Without observing the dimensions of the patch (Sokabe et al. 1991) and knowing the mechanical history of the patch, one cannot be sure of the origin of changes in activation following any treatment.

Regardless of the difficulty in quantifying the sharing of tension among cortical elements, the status of cytoskeletal linkages seems to be critical in MSC function. Confirmation that an intact cortical cytoskeleton can buffer MSCs from mechanical stimuli comes from work with snail neuron stretch-activated K^+ channels (Small and Morris 1994)and from stretch-activated Cl^- channels in kidney cells (Schwiebert et al. 1994). In the latter case, it seems plausible that actin-polymerization can attenuate the mechanosensitive response of the channels.

It should be emphasized that the stresses involved in seal formation of a patch can influence not only MSCs, but other channels including traditional voltage-sensitive Na^+ channels (Fahlke and Rudel 1992) as well.

4.6.1
Gentle Patches

In molluscan neurons and in *Xenopus* oocytes, activation of SACs following a step stimulus of suction has been studied in "gentle" patches, i.e., ones formed using minimal mechanical disruption (Small and Morris 1994; Hamill and McBride 1992). A cartoon depicting expected patch structure before and after mechanical disruption produced by "nongentle" stimulation is shown in Fig. 12. The responses of MSCs in the gentle patches differ from those in disrupted patches. Gentle patches from oocytes show rapid adaptation while snail neuron channels show delayed activation. The existence of delay in molluscan neurons and of adaptation in *Xenopus* oocytes that have not been subjected to disruptive mechanical stimuli cannot be explained by possible smaller patch size associated with nongentle patches.

In gentle patches made on molluscan neurons, SA channel activation in response to a "first hit" (a relatively large step of applied suction) proceeds only after a delay. The first hit delay is sufficiently long – >2 s at –130 mmHg – that decreases in delay produced by various treatments are easily detected. In fact, delay decreases steeply whenever cortical cytoskeleton integrity is compromised. Treatments that decrease delay include repetition of mechanical stimulation, use of larger mechanical stimuli, pretreatment with cytochalasin (drugs that promotes actin depolymerization), use of recently isolated (i.e., recently disrupted) cultured neurons, hyposmotic swelling of neurons, exposure to the sulfhydryl re-

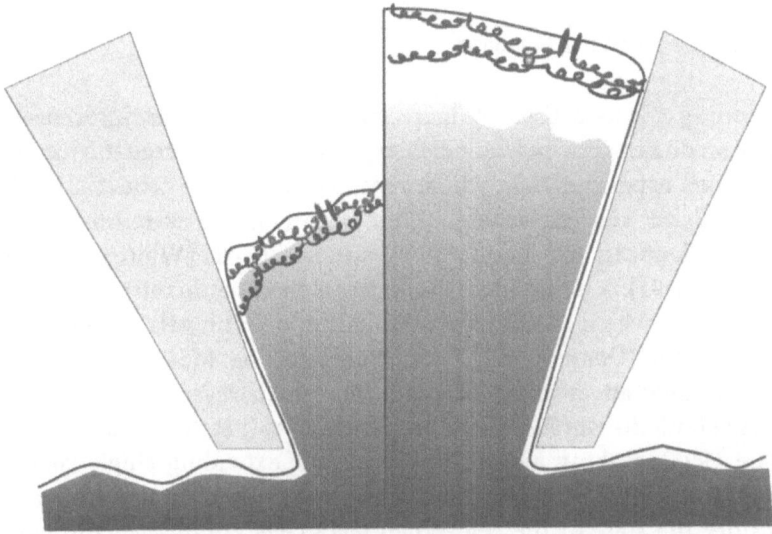

Fig. 12. Interpretation of a "gentle" patch (*left*) and a repeatedly exercised patch (*right*). The membrane skeleton is shown as the curly strands near the top and the interior cytoskeleton as the shaded region. An MSC is shown attached to the membrane skeleton

agent, N-ethylmaleimide, and elevation of intracellular Ca^{2+} (Small and Morris 1994; Morris 1996). In addition to a shorter delay, these treatments also lead to increased levels of activity of the channels for the same pressure. Thus, in two ways, speed of response to mechanical stimulation and extent of activation; an intact cortical cytoskeleton makes it more difficult to activate these channels. If considerably smaller first-hit pressures are used (say, –20 to –50 mmHg) on gentle patches, delays of tens of seconds are readily obtained, even though in "primed" snail neuron patches (exercised "nongentle patches"), channels are routinely activated by –20 mmHg, for example (Small and Morris 1995, 1995a). Delayed activation of SACs ($\approx$30 s) has also been reported by Kim and colleagues (1993). Although the volume-activated Cl channel in lymphocytes has not been confirmed as a stretch-activated channel it is notable that comparable delays occur during activation of this conductance mechanism (Lewis et al. 1993). It is possible, however, that this delay is caused by the need to integrate the osmotic water flux to produce a change in volume.

4.6.2
Whole-Cell Time-Dependent Responses

Confounding effects caused by increasing patch size following repeated stimulation do not arise in whole-cell experiments and there are a number of whole-cell experiments in which repeated mechanical stimulation increases cellular responsiveness. These experiments were done with osteoblasts (Duncan and Hruska 1994) and heart cells (White et al. 1993; Sasaki et al. 1992). A consistent hypothesis is that the thixotropic properties of actin, in which mechanical stimulation of actin gels promotes the sol-state of actin (Dennerll et al. 1988), may affect net MSC sensitivity.

There is now an example of a neuron which reveals its MS cation currents at both the whole-cell and single-channel (patch) levels (Oliet and Bourque 1996), making a good prototype for extending single-channel observations. Where macroscopic currents are detectable, it is important to consider the state of the cell. When GH3 cells are inflated through a whole-cell pipette, MS currents only appear after irreversible inflation is effected (Morris and Horn 1991b), suggesting that cortical disruption may enhance access of mechanical stimuli to the channels. Failure to disrupt the cortex may explain why, in snail neurons (Wan et al. 1995), application of pressure via a whole-cell pipette seldom yields MS current, but prolonged neuronal inflation, engendered by osmotic swelling, does induce current (Wan et al. 1995). Perhaps disruption is more extensive with osmotic swelling.

In patches from chick skeletal muscle, Sachs and coworkers (Sachs 1987; Sokabe et al. 1991) found properties that would be consistent with this idea. A step of pressure led to a slow ($\approx$0.5 s) increase in patch area and channel activity. That is, although there were changes in local stress, the stress seemed to be shunted across the channels. Sustained deformation, however, did produce responses – this is evidence of an effective viscosity. Because of the compliance of the parallel and series elements of the cytoskeleton, MSCs in situ may behave as detectors of macroscopic strain rather than stress. As implied in Fig. 12, changes in patch size and changes in cortical integrity are correlated, so one can expect the interpretation of input/output relations for the channels to be fraught with difficulty.

4.6.3
Effects of Repeated Stimulation

Delayed activation is not universal. Hamill and McBride (1992) showed that in *Xenopus* oocytes, the MSC response to a step of pressure is close to being limited by the rise time of the pressure. The pressure stepping apparatus used by Small and Morris (1994) for both oocytes and snail neurons had a slower rise time, but, like Hamill and McBride (1992), they obtained immediate activation in *Xenopus* oocytes. A similarity of these two very different preparations is that repeated stimulation effectively abolishes the dynamic aspects of the response. In neurons, the delay is lost, and in *Xenopus* oocytes, the rapidly adapting part of the response (Hamill and McBride 1992) is lost. In "traumatized" patches from either preparation, a step stimulus leads immediately to a sustained increase in channel activity. Overall, this suggests, not surprisingly, that in *intact* oocytes and *intact* neurons, cytoskeletal arrangements that affect channel behavior can differ.

It is important to keep the dynamic and fragile responses of MSCs in mind when interpreting data. For example, if a 1 s pulse of −50 mmHg were applied to a gentle patch on a snail neuron, one might conclude that the neurons lacked SACs. Longer stimulation would, however, reveal channel activity. In the opposite case, a phasically responding system (discussed below) would not exhibit channel activity unless the pressure were applied rapidly. Moreover, where the phasic response is mechanically fragile, it would be easy to observe only the small tonic fraction of the response and therefore overlook possible physiological roles for the MSCs involved.

Phasic responses are seen in cells as diverse as tunicate (Moody and Bosma 1989) and amphibian oocytes (Hamill and McBride 1992), as well as yeast (Gustin 1992) and bacterial MSCs (Hase et al. 1995). In no case do we understand the role of these channels. Although it seems likely that the cytoskeleton is implicated in these cellular systems, the observation of adapting responses by bacterial MscL channels reconstituted into liposomes (Hase et al. 1995) raises the possibility of alternate modes of adaptation involving only the channel and the bilayer (see Sect. 2).

The fragility of MSC dynamic responses opens the possibility for substantial modification of MSC responses by agents, particularly hormones, capable of modifying the cytoskeleton. For example, vasopressin causes actin depolymerization and SAC activation in kidney cells (Verrey et al. 1995). During the cell cycle there are major cytoskeletal rearrangements that may account for the observations (Medina and Bregestovski 1991; Bregestovski et al. 1992) that SAC sensitivity varies during the cycle.

Large variations in mechanosensitivity that are dependent on varying cytoskeletal structure may also account for apparent variations in the density of SACs toward hyphal tips in fungi (Levina et al. 1994). Although the authors suggest, quite plausibly, that actin causes the channels to cluster, either by linking channels to the cytoskeleton or by increasing transport of channels to the tip, regional variations in effective patch size and/or compliance could explain the data. In other words, channels may be present at uniform density but they may be difficult to stimulate mechanically in "old" membrane far from the tip and easy to stimulate in "new" membrane near the tip. This echoes the results for molluscan neurons in which SACs become more and more difficult to activate with mechanical stimuli as the neurons become better established in culture (Small and Morris 1994).

The cytoskeleton also seems to affect weakly mechanosensitive ligand-gated channels. The activity of NMDA channels can be potentiated by pipette suction in outside-out patches, inside-out patches and cell-attached patches (Paoletti and Ascher 1994). By contrast, kainate-activated glutamate channels in the same patches are not mechanosensitive. The potentiation by pressure is not large compared to many MSCs – threefold increases in activity with about ±80 mbar (±65 mmHg). However, successive stimuli yield progressively larger effects, suggesting that the receptor sensitivity of NMDA channels to stretch might be increased by disrupting the cytoskeleton. As mentioned earlier, NMDA responses to stretch in the whole-cell mode tend to support this view, since osmosensitivity only became evident after prolonged dialysis. The experiments were performed in the presence of internal pH and pCa buffers, without ATP or GTP, and in the presence of F^- (which blocks most phosphatases), so that the "run-up" of mechanosensitivity is likely related to improved transfer of force to the NMDA channels. Confounding this interpretation, however, Rosenmund and Westbrook (1993) reported that cytochalasins produce NMDA current *rundown* while phalloidin stabilizes the response.

4.7
Adaptation of Mechanosensitive Channels

As mentioned above, SACs from widely varying cell types exhibit inactivation or adaptation in the face of sustained steps of applied suction. The

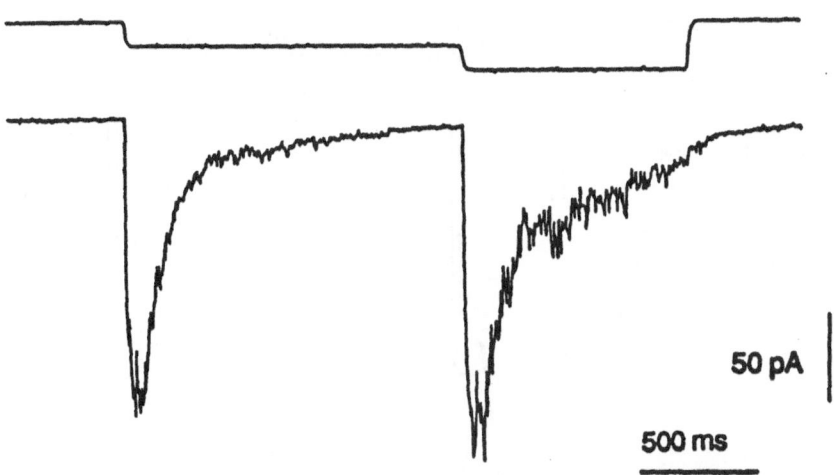

Fig. 13. Response of *Xenopus* oocyte MSCs to a two-step staircase of pressure. *Upper trace,* two steps of suction, 8 mmHg each; *lower trace,* the MSC current (produced by many channels). (From Hamill and McBride 1992, with permission)

phenomenon was noted in yeast (Gustin et al. 1988; Gustin 1992), tunicate eggs (Moody and Bosma 1989), and most recently in *Xenopus* oocytes (Hamill and McBride 1992). For the *Xenopus* channels, a pressure servo (McBride and Hamill 1992) or switched pressure source (Small and Morris 1994) were used for making the step stimuli. Figure 13 shows an example of adaptation. There are a number of intriguing results from this work, but it is important to realize that the results do not the represent typical properties of most MSCs.

The first interesting feature is that with to a two-step staircase of pressure, the peak responses are the same amplitude, and both responses decay to near the baseline – there is no steady state activation nor any accumulated inactivation! In contrast, were this a classical Na^+ channel, the inactivation produced by the first step would reduce the peak response to the second step. This kind of "recovered" inactivation is called adaptation. It is seen in other channels, notably the Ca^{2+} release channels of intracellular stores (Gyorke and Fill 1993). There are two basic ways to explain this phenomenon: adaptive state kinetics (Sachs et al. 1995) and viscous coupling. The kinetic model explanation, is shown in Fig. 14.

The adaptive behavior can be understood by examining the extreme cases, tensions T=0 and T=∞. When T=0, k_{23} and k_{14}=0 and the channel

Fig. 14. A state model for adaptation. C_1 and C_3 refers to closed states and O_2 and O_4 to open states. T is tension. In detailed balance, this model can produce exact adaptation where the steady state response is independent of T and only transient changes occur. (From Sachs et al. 1995)

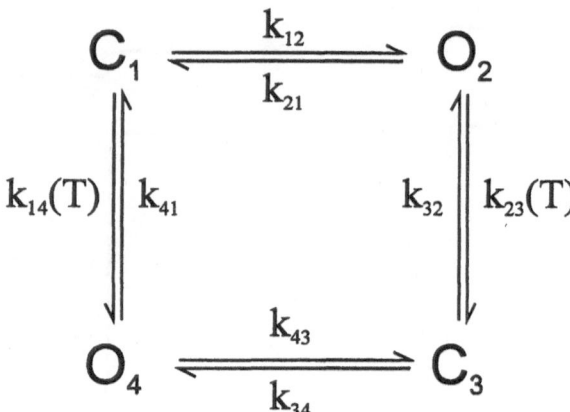

oscillates between states 1 and 2 establishing the resting probability of being open. When $T=\infty$, k_{23} and $k_{14}=\infty$ and the channel oscillates between states 3 and 4 establishing the saturated probability of being open. If $k_{12}/k_{21}=k_{34}/k_{43}$ then the probability of being open in the two cases is equal and there is no steady state response. The response to a step can be seen if one assumes that at rest most channels are in C_1. When a step of T is applied, channels move from state 1 to 4, producing current. The channels in open state 4 then slowly decay to closed state 3, causing adaptation.

In attempting to fit the model in Fig. 14 to Hamill and McBride's (1992) staircase data, quantitative conflicts appear (Sachs, unpublished). If the probability of being open at rest is small, as appears to be the case, then obtaining equal sized peaks requires that each step only alter the population of C_1 only slightly. This means that the peak currents represent activity of only a small fraction of the available channels so that the channel density would have to be orders of magnitude higher than that estimated from steady state experiments. While this conclusion is possible, it seems unlikely. With repeated stimulation of oocyte patches, adaptation is lost but the channels show steady state activation (see Yang and Sachs 1990; Methfessel et al. 1986) that is *smaller* than the peak currents seen when adaptation is intact (Hamill and McBride 1992). Changes can be made to the model of Fig. 14 by adding states, but no state model that is in detailed balance will allow a staircase stimulus to produce equal sized responses in more than two steps. Experiments using multiple steps would prove very useful.

Fig. 15. Mechanical equivalent of viscoelastic adaptation. Sudden stretches apply large forces to the channel. With time, relaxation of the dashpot (viscosity) removes the component of force applied to the channel

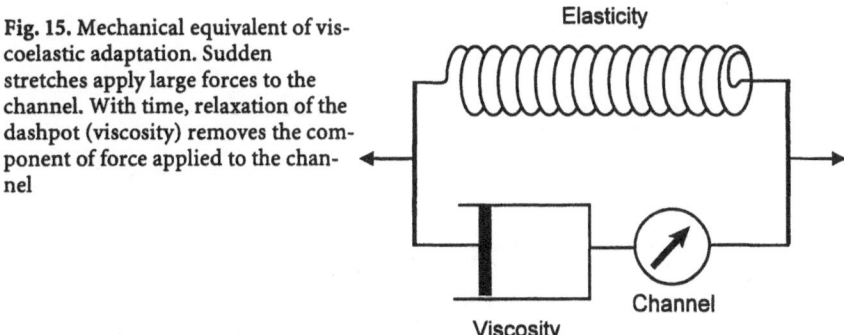

An alternative explanation for adaptation is that it arises from relaxation of the stimulus through a viscoelastic coupling as shown in Fig. 15. This model has the advantage of allowing a large number of equal amplitude peak responses in a staircase. Involvement of the cytoskeleton is also suggested by the observation that repeated stimulation can disrupt adaptation without blocking activation (Hamill and McBride 1992). However, additional data on the voltage dependence of the adaptation makes strict cytoskeletal dominance of adaptation unlikely.

Figure 16 shows that adaptation is present only at hyperpolarized potentials. Since the cytoskeleton is outside the membrane field, the voltage dependency of adaptation for a cytoskeletal model would imply a voltage dependent viscosity. This would have to arise from interaction of the cytoskeleton with charged intramembrane components. The charges must be significant since the time constant of adaptation changes from ≈50 ms at −100 mV to ≈0 at +100 mV. Highly voltage dependent membrane motors have been observed in outer hair cells of the cochlea, but these seem to be associated with elaborate and organized internal structures (Holley and Ashmore 1990).

It is possible to explain the basic features of voltage dependent adaptation with the adapting kinetic model of Fig. 14 by allowing the horizontal reactions to be voltage sensitive. For example, if k_{43} were to decrease and k_{12} increase with depolarization, adaptation rates and the magnitude of the peak currents would be reduced. It remains to be seen whether the data can be accounted for quantitatively. It is possible that the channel is not in detailed balance; i.e., it is using energy from the electrochemical flux. This flux is not Ca^{2+} since adaptation appears to be Ca^{2+} independent (Hamill

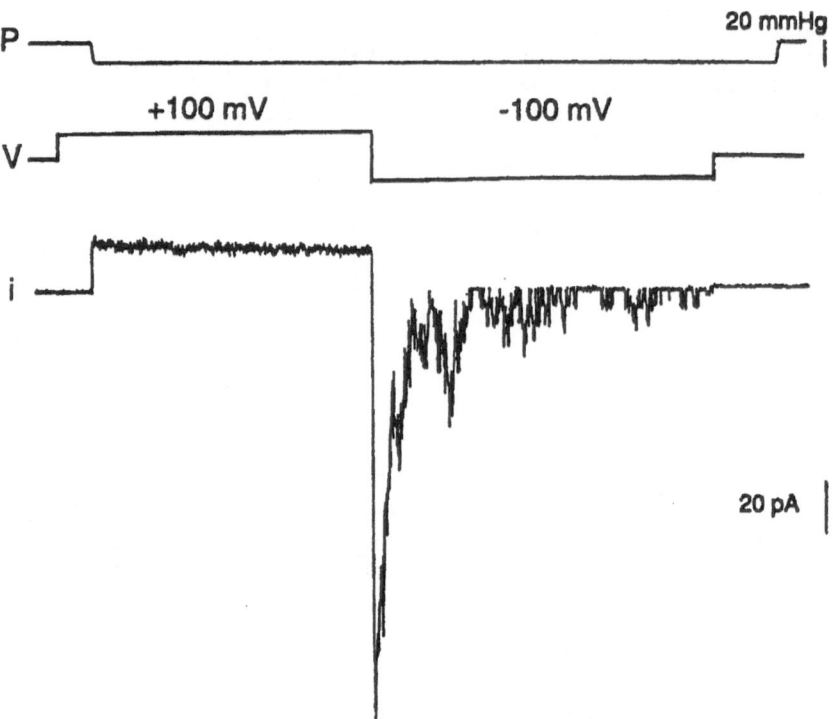

Fig. 16. The effect of membrane potential on adaptation in *Xenopus* oocytes. (From Hamill and McBride 1992, with permission)

and McBride 1992). If the channel is not in detailed balance, its rate of adaptation should be affected by the direction of the current at constant voltage; this is testable by varying the ionic gradients.

One might imagine that the ideal test for the origin of adaptation would be the MscL channel reconstituted in a bilayer, since the cytoskeletal mechanics are eliminated. Martinac and coworkers (Hase et al. 1995) recently demonstrated that MscL makes transient responses to tension in a bilayer! Unfortunately, they used only single step stimulation and did not examine the voltage dependence, so it unknown whether the response represents inactivation or adaptation. The basic response, however, is remarkably similar to that seen in *Xenopus*. As discussed in Sect. 2, the MscL adaptation may be caused by viscous relaxation associated with slippage of one monolayer over the other.

4.8
Physiological Role of Adaptation

The possible physiological significance of adaptation, as seen at the single-channel level, remains to be determined. It has a clear role in the phasic behavior of specialized mechanoreceptors such as the muscle spindle and the Pacinian corpuscle, where it emphasizes changes in the stimulus (Munger and Ide 1987; Akoev et al. 1988). At the cellular level, output adaptation may have nothing to do with the cytoskeleton or MSCs. Where it has been studied in mechanosensitive cells, adaptation is mediated by activation of K^+ currents (French 1986; Swerup 1983; Erxleben 1993; Wellner and Isenberg 1995). A case in point is urinary bladder myocytes: Studies with the whole-cell clamp suggest that SACs mediate stretch-induced inward currents (Wellner and Isenberg 1995). Myocyte stretch during rapid bladder filling would thus increase the myocyte's level of spontaneous electrical activity. In the face of continued extension, there would be a slow fall in muscle tone and a relaxation of the bladder. Why Mother Nature would think this behavior is good is unclear to us! In a single-channel patch clamp study of myocyte SACs (Wellner and Isenberg 1995), no evidence was found for adaptation, but it is possible that the stresses of patch formation destroyed adaptation (Hamill et al. 1992; Small and Morris 1994). The authors suggested that adaptation was a secondary effect of SAC activation: Ca^{2+} influx activating large conductance Ca^{2+} activated K (BK) channels. The Ca^{2+} sensitivity of BK channels and the stretch sensitivity of the cationic SACs were both enhanced by cAMP.

It is difficult to hypothesize about evolutionary advantages of adaptation in an undifferentiated cell like the oocyte, but it is easy to imagine the advantages of the more common "low-pass filtering" effects (Sokabe et al. 1991; Small and Morris 1994). A muscle or bone cell may need to be informed of sustained mechanical inputs in order to modify some aspect of its metabolism (volume regulation, Ca^{2+} loading, triggering of protein synthesis), but to have it subjected to large electrical inputs when momentarily stressed may be counterproductive.

5
Pharmacology of MSCs

This area has proved to be a nagging problem. There are no specific drugs available to modify MSCs. Progress is beginning to be made in this area, however, as drug profiles are developed for the channels. There are many reports of blockers (Table 3), but few reports about channel activators (Table 4).

Table 3. Effects of blocking drugs on MSCs and other mechanical processes

Drug	IC$_{50}$ (μM)	Preparation	Reference
Gd^{3+}	5	*Xenopus* oocyte, SAC Cat	Yang and Sachs (1989)
	≈5	*Xenopus* myocyte, SAC	Yang and Sachs (1989)
	100, ne	*Lymnea* neuron, SAC K$^+$	Small and Morris (1995)
	??	*mdx* mouse SIC Cat	Franco et al. (1991)
	≈500	*E. coli*	Hase et al. (1995), Berrier et al. (1992)
		Dog heart arrhythmias	Hansen et al. (1991)
Amiloride	500	*Xenopus* oocyte, SAC Cat	Lane et al. (1991)
	2000	*Lymnea* neuron, SAC K$^+$	Small and Morris (1995)
	53	Mouse hair cell,[a] Cat	Rusch et al. (1994)
	50	Chick hair cell,[a] Cat	Jorgensen and Ohmori (1988)
BrHMA	34	*Xenopus* oocyte, SAC Cat	Lane et al. (1992)
TEA	5000	*Lymnea* neuron, SAC K$^+$	Small and Morris (1995)
Quinidine	750	*Lymnea* neuron, SAC K$^+$	Small and Morris (1995)
Diltiazem	100, ne	*Lymnea* neuron, SAC K$^+$	Small and Morris (1995)
	5	Chick heart, SAC Cat	Ruknudin et al. (1993)
Streptomycin	<40	Guinea pig heart, [Ca]$_i$	Gannier et al. (1994)
	20	Chick hair cell[a]	Kimitsuki and Ohmori (1993)
	23	Chick skeletal muscle SAC Cat	Sokabe et al. (1993a)
	80	Guinea pig heart arrhythmias	Nazir (1994)
	2700	Mouse skeletal muscle SAC Cat	Winegar et al. (1996)
Penicillin	<40	Guinea pig heart, [Ca]$_i$	Gannier et al. (1994)
Kanamycin	10	Chick skeletal muscle SAC Cat	Sokabe et al. (1993a)
	≈4000	Mouse skeletal muscle SAC Cat	Winegar et al. (1996)

Table 3. Continued.

Drug	IC$_{50}$ (μM)	Preparation	Reference
Neomycin	2.4	Chick skeletal muscle SAC Cat	Sokabe et al. (1993a)
	≈400	Mouse skeletal muscle SAC Cat	Winegar et al. (1996)
Ribostamycin	≈40	Chick skeletal muscle SAC Cat	Sokabe et al. (1993a)
Dibekacin	≈30	Chick skeletal muscle SAC Cat	Sokabe et al. (1993a)
Dihydro-streptomycin	≈400	Mouse skeletal muscle SAC Cat	Winegar et al. (1996)
Gentamycin	≈1400	Mouse skeletal muscle SAC Cat	Winegar et al. (1996)
Amikacin	≈2600	Mouse skeletal muscle SAC Cat	Winegar et al. (1996)
TTX	≈5	Chick heart, SAC Cat	Ruknudin et al. (1993)
TEA	5000, ne	Crayfish, SAC Cat	Ruknudin et al. (1993)
4AP	1000, ne	Crayfish, SAC Cat	Ruknudin et al. (1993)
Ethanol	0.65 M (3%), ne	*Lymnea* neuron, SAC K$^+$	Small and Morris (1995)
GS venom	≈1:1000	GH3 SAC Cat, [Ca^{2+}]$_i$	Chen et al. (1996)
	≈1:1000	*Xenopus* oocyte, chick heart[a]	Niggel et al. (1996)

BrHMA, Bromohexamethyleneamiloride; TEA, tetraethyl ammonium; 4AP, 4-aminopyridine; GS, *Grammostola spatulata*; ne, no effect; [Ca^{2+}]$_i$., intracellular Ca^{2+}.
[a] Mean current.

Table 4. Effects of activating drugs on MSCs

Drug	≈IC$_{50}$ (μM)	Preparation	Reference
Pinacidil	≈10	Rat heart SAC K$^+$ (*KATP*)	Van Wagoner (1993)
Chloropromazine	≈20	*E. coli*	Martinac et al. (1990)
Chloropromazine	≈10	Chick skeletal muscle	Sokabe et al. (1993a)
Trinitrophenol	≈500	*E. coli*	Martinac et al. (1990)

5.1
Gadolinium

The most common blocker currently used is the rare earth, gadolinium (Yang and Sachs 1989). This trivalent ion acts like La^{3+}, but lower doses are required. Gd^{3+} blocks MSCs in a host of different preparations from MscL of *E. coli* (Berrier et al. 1992), to human neurons (Quasthoff 1994). Although it is effective at blocking most MSCs at concentrations below 100 μM, it also blocks other channels, including some voltage dependent Ca^{2+} channels (Lacampagne et al. 1994; Docherty 1988) and colicins (Bonhivers et al. 1995). K^+ selective MSCs have proven to be more resistant than others (Small and Morris 1995). Aside from its lack of specificity, Gd^{3+} also has the problem of precipitating physiological anions such as HCO_3^- and PO_4^{-2} and therefore needs to be used with more inert buffers such as HEPES. Despite these problems, Gd^{3+} does block many MSCs. It has even been used to block mechanically induced responses in the whole heart without blocking the normal beat or contraction (Hansen et al. 1991) as well as gravitropism in *Euglena* (Lebert and Hader 1996). Gd^{3+} sensitivity has been used as a test for involvement of MSCs in stretch sensitive processes such as proliferation of lungs cells (Liu Xu et al. 1994), pressure volume relationships (Takano and Glantz 1995) and the length tension relationship (Lab et al. 1994) in the heart, activation of phospholipase C (Matsumoto et al. 1995), modulation of channels activated by antidiuretic hormone (Marunaka et al. 1994), and many more.

We want to reemphasize that sensitivity to Gd^{3+} does not serve as a reliable marker for physiological MSC activity! It has a wide spectrum of cross reactivity. The blocking effect seems to consist of at least two components: an open channel block, and some more complex higher order response (denaturation?) at higher concentrations (Yang and Sachs 1989). Data on MscL show that Gd^{3+} shifts the activation curve to higher tensions (Sukharev et al. 1993).

The broad spectrum of Gd^{3+} activity on MSCs is peculiar. It is surprising that a single agent is active against a 3 nS bacterial channel and a 25 pS eucaryotic channel. Gd^{3+} has the crystal radius of Ca^{2+} but is trivalent and hence may be effective in pulling together negatively charged lipid head groups. It is possible that the action of Gd^{3+} is not on the channel protein at all, but on the surrounding lipids, pulling them tightly together so the membrane surrounding the channel is rigid or somehow unable to support a change in channel dimensions. It would be useful to explore the effect of Gd^{3+} on the elastic properties of phospholipid membranes (Evans and Needham 1987) and on channel-lipid interactions that might mimic dimensional changes of MSCs, using alamethicin (Opsahl and Webb 1994a) or gramicidin (Elliott et al. 1983). Preliminary data on the effects of Gd^{3+} on phospholipid membranes indicates that Gd^{3+} induces major changes in the phase transition temperature and a large change in dipole potential, both effects consistent with an MSC inhibition originating in the lipids (Ermakov et al. 1996). It is possible that those MSCs that are insensitive to Gd^{3+}, for example, the molluscan SA K channel (Small and Morris 1995), confine the dimensional changes of their mechanosensors to internal regions of the channel structure where space has already been provided.

5.2
Amiloride

Amiloride and its derivatives have been used to block MSCs in different systems. In the millimolar range amiloride blocks the MSCs of cochlear hair cells (Jorgensen and Ohmori 1988; Rusch et al. 1994), cationic SACs in the *Xenopus* oocyte (Lane et al. 1991, 1993; Hamill et al. 1992), and K^+ selective SACs in molluscan neurons (Small and Morris 1995). Block by amiloride has a Hill coefficient of 1.5–1.8, suggesting that more than one molecule is associated with block; the block also exhibits a slight voltage sensitivity. Some derivatives of amiloride are more efficacious than the parent compound, and the variation of efficacy with structure has been suggested as a tool to identify responses arising from MSCs (Hamill et al. 1992). Amiloride is clearly not selective since it blocks a host of transporters, including the Ca/Na exchanger and the Na-H exchanger, some of which are blocked in the nanomolar range (Benos et al. 1992). Like gadolinium, amiloride is more useful as a biophysical probe of MSCs than as a physiological tool.

5.3
Antibiotics

Many cationic antibiotics block mechanical transduction (see Table 4) but they also block other channels, including Ca^{2+} channels (Haws et al. 1996). In a recent analysis of SACs in muscle, a number of the antibiotics produced subconductance states, as though they did not block the channels directly but did effect partial block for permeant ions (Winegar et al. 1996). Given that these agents are charged, it is not surprising that the block by these drugs is voltage dependent, with the apparent binding sites located between 10 and 50% of the way through the membrane field (Winegar et al. 1996). The affinity of various antibiotics for the SACs in mouse skeletal muscle was: dihydrostreptomycin, neomycin >gentamycin amikacin, streptomycin >kanamycin. The block by antibiotics is complicated by a stoichiometry of >1, so that the apparent affinity is concentration dependent (Winegar et al. 1996). It should also be noted, when comparing experiments, that competition with Ca^{2+} can reduce the effective affinity for the antibiotics.

5.4
Peptides

A spider venom has been identified that contains a peptide blocker effective for a number of SACs including those in GH3 cells, *Xenopus* oocytes and chick heart cells (Chen et al. 1996; Niggel et al. 1996). Venom isolated from the spider, *Grammostola spatulata* (GS), blocks the Ca^{2+} uptake associated with hypotonic swelling, but not the voltage dependent L-type Ca^{2+} channels. The raw venom doesn't block action potentials in the guinea pig heart, but does block stretch induced changes in action potential configuration (Nazir et al. 1995). Interestingly, when GS venom is added to GH3 cells in normal saline, Ca_i^{2+} drops to levels equal to those observed in the absence of extracellular Ca^{2+}. This suggests that, at least in the GH3 neurons (Chen et al. 1996), SACs may be active at rest (not necessarily via resting stretch, but utilizing stretch independent rates), and that they represent a primary source of resting Ca^{2+} uptake. The availability of a peptide blocker raises the possibility of histochemistry and of tracking MSCs for cloning.

5.5
Activators

No clear activators are known to act on the channels directly. The amphipaths, as discussed in Sect. 2 and referenced in Table 4, probably do not act directly on the channels, but on the surrounding lipids.

5.6
Cautions

Many drugs can alter the sensitivity of MSCs by affecting the mechanical coupling to the channels. Actin reagents are one example. Any attempt to characterize a pharmacological agent as affecting intrinsic sensitivity requires extremely careful micromechanics to make the result convincing. Changing the culture or isolation conditions of cells may have the same effects (e.g., Small and Morris 1994), and can easily cause apparent changes in channel density by changing the size of patches.

6
Physiology of MSCs

6.1
Stretch-Inactivated Cation Channels:
The Basis for Osmotransduction by Hypothalamic Neurons

A variety of MSC currents are evident at both the single channel and macroscopic levels, but determining the contribution of these currents to the physiology of the cell remains difficult. Recently, this "physiological barrier" has been convincingly breached for a specialized mechanoreceptor cell (also, see the section on the heart, below, for an example of MSC activity correlated with physiological in a non sensory cell). Unlike hair cells of the acoustico-lateralis system, however, the mechanoreceptor in question has no highly specialized morphology that one can associate with mechanotransduction. Called magnocellular neuron cells (MNCs), they constitute a subpopulation of neurons in the hypothalamus that are responsible for detecting and responding to blood osmolarity. They possess mechanosensitive cation channels that appear to underlie osmotransduction (Oliet and Bourque 1993, 1994; Bourque et al. 1994). These currents hyperpolarize the neurons during hypoosmotic stimuli and depolarize them during hyperosmotic stimuli, as summarized in Fig. 17. This in turn leads to respective decreases and increases in release of the osmoregulatory hormone, vasopressin (Oliet and Bourque 1993; Bourque et al. 1994). Under cell-attached recording conditions with physiological solutions, the MNC channels have a conductance in the 30 pS range and reverse near

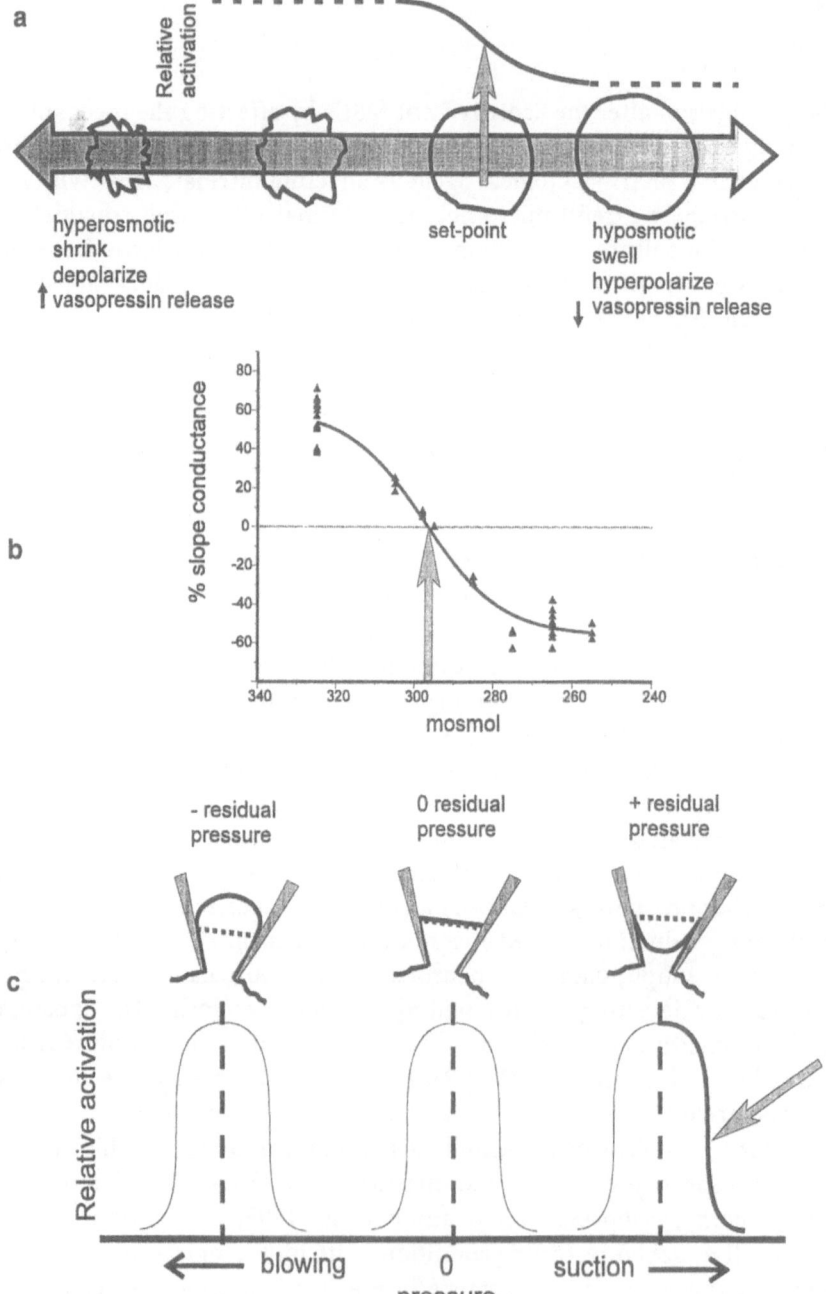

−40 mV, this is sufficiently depolarized with respect to the cells' resting potential for active channels to be excitatory.

Activation curves for the MNC mechanosensitive single-channel currents in cell-attached patches are unusual. Rather than the expected sigmoid curve, the MNC patches invariably yielded a bell-shaped activation curve. This may stem from artifacts in the pressure recording and/or renormalization of the data by the authors. Although appearing to represent positive and negative applied pressures, the activation curves were actually obtained only at negative applied pressures (suction), a point not immediately evident because the authors offset their data so that the peak occurred at the origin (Oliet and Bourque 1993). In contrast to the single-channel data, the macroscopic currents showed a monotonic dependence on hypotonicity. If the microscopic and macroscopic recordings reflect activity of the same channels, the difference requires explanation.

A straightforward explanation for the bell curve emerges from two assumptions: (a) the channel is operationally a stretch-inactivated channel and (b) the channel activity is recorded under conditions of residual positive pressure in the pipette. A bell-shaped activation curve is expected when the independent variable is pressure, but not when it is tension (Morris and Sigurdson 1989). Since the experimental x-axis refers to the applied pressure rather than membrane tension, the zero tension point is not truly known.

The bell-shape can occur because the gigaohm seal attaches membrane firmly to the electrode wall allowing the patch to bow upward or downward, depending on the effective pipette pressure. If residual pressure

---

Fig. 17A–C. MSCs in osmosensory neurons with bell-shaped and sigmoid activation curves. *A Above*, the sigmoidal activation curve for a SIC, where hyperosmotic conditions (low tension) are shown *at left* and hypoosmotic conditions (high tension) *at right*. The sigmoid's slope is negative, corresponding to data obtained from MNCs. *Vertical arrows* (A,B), the physiological set point for blood osmolarity. B Whole-cell slope conductances (percentage of control) for the osmosensitive current, as a function of bath osmolarity. (From Oliet and Bourque (1993) with permission; we have fit a Boltzmann to the data) C Single-channel "bell-shaped" activation curves observed with different resting pressures in the pipette (or possibly cytoskeletal stresses). *Dashed line*, membrane position for minimal membrane tension and maximal SIC activity; *solid line*, its position at zero *applied* pressure. When the two do not coincide (e.g., when residual pressure in the pipette provides an offset), the response curve is displaced. *Bold in right curve* corresponds to the sigmoids observed in A and B

persists in the pipette after gigaseal formation, the bell-shaped curve will be centered not at the origin, but elsewhere along the axis of experimentally applied pressure, according to the sign and magnitude of the residual pressure (Fig. 17). To summarize, the bell-shape probably arises from errors in setting the proper pressure (the "zero" pressure) to obtain minimal stress in patches.

We would like to reemphasize, here, that there is a need to properly adjust the pipette pressure when doing experiments on MSCs. The best way to assure an accurate zero pressure in the pipette is to monitor the flow prior to forming the seal. By watching the movement of small particles in the bath, it is easy to determine quite accurately what recorded pressure corresponds to true zero pressure. This calibration removes all influences of residual capillarity and possible resting pressure offsets in the measuring system. It does not, however, remove the influence of cytoskeletal forces pulling on the membrane, an effect that can only be tested by direct observation of the patch (Sokabe and Sachs 1990).

Considering the possible role of pressure offsets, it is understandable that SICs may yield bell-shaped activation curves, although, one would also have to make the assumption that all the records were obtained with positive resting pressure in the pipette. The prediction then is that the activation curve of macroscopic SI currents obtained without offset is sigmoidal. The published data were fit to a straight line not a sigmoid curve, but as we show below (Fig. 17B) the data are quite compatible with a sigmoid curve, although data covering a somewhat wider range of osmotic pressures are needed.

The idea that SICs give rise to the osmotransduction currents is supported by the demonstration that Gd^{3+}, a common blocker of many MSCs, blocks both the macroscopic osmosensor currents and the single-channel SI currents (Oliet and Bourque 1996). The pressing issues now are to understand what it is about these particular cation channels that makes them highly sensitive to the mechanical stimuli associated with small osmotic perturbations. Do the neurons have cortical skeletons with few parallel elastic elements? Is their large size important to their osmosensory role as it might be if they are governed by Laplace's law (detecting osmotic pressure gradients as membrane tension)? Do the neurons have an unusually high water permeability? What is the basis of the "offset" – is it really a positive pressure in the pipette? Is there something special about the way the membrane interacts with the pipette, so that at zero pressure the patch is pulled towards the cytoplasm? While there are many questions, all of them seem to be answerable with current technology.

6.2
Other Correlations of Whole-Cell Activity and MSCs

6.2.1
Smooth Muscle

Smooth muscle is capable of autoregulation; i.e., in response to stretch, it can actively contract and tend to maintain a constant volume of its organ or to simulate peristalsis (Johansson and Mellander 1975). Some of this activity appears to arise from smooth muscle cells directly (Hisada et al. 1991; Kirber et al. 1988; Wellner and Isenberg 1994; Davis et al. 1992), and some is coupled from endothelial cells (Goligorsky 1988; Hoyer et al. 1994). In arterial smooth muscle cells, Davis and colleagues. (Davis et al. 1992) recorded the activity of cationic SACs using cell-attached recording, they demonstrated that stretching the cells between two suction pipettes caused a whole-cell current that reversed at ≈-15 mV. Cell currents of tens of pA were elicited for $\approx 15\%$ increases in length; and this was considered capable of accounting for autoregulation of vessel tone. Unfortunately Davis and colleagues did not show that the whole-cell and single-channel currents were pharmacologically similar. Results similar to those of Davis and colleagues have been obtained for urinary bladder smooth muscle cells by Wellner and Isenberg (Wellner and Isenberg 1994, 1995).

6.2.2
Heart Muscle

The heart is sensitive to mechanical deformation. Characteristically, the beat rate increases with atrial pressure (Bainbridge 1915). The early embryonic chick heart, while still a quiescent tube, can be stimulated to begin beating by elevating the internal pressure (Rajala et al. 1976, 1977). The whole heart, when stretched with an intraventricular balloon, will generate extra systoles in response to stretch (Lab 1980; Hansen et al. 1990a; Stacy et al. 1992). It has been postulated that many of the fatal arrhythmias that follow a heart attack are caused by stretch of the weak tissue around the infarct which then generates excitatory currents (Hansen et al. 1990b). In the dog heart, stretch-induced systoles are blocked by Gd^{3+} but not by Ca^{2+}

channel blockers, which suggests a role for SACs (Hansen et al. 1990b). SACs have been reported in heart cells from rat, guinea pig, chick and snail (Sigurdson et al. 1987; Sadoshima et al. 1992; Kim 1992; Sasaki et al. 1992; Akay and Craelius 1993; Craelius 1993; Ruknudin et al. 1993; Van Wagoner 1993), but the reported selectivity and gating responses vary widely; thus it is difficult to make general predictions about the functions of these channels.

The only whole-cell voltage clamp experiments on the effect of mechanical strain were done by Sasaki et al. (1992) in rat ventricular cells and by Hu and Sachs (1994, 1995, 1996) in chick heart cells. Both groups found cationic, mechanosensitive, currents that reversed at about –15 mV. Hu and Sachs (1996) were able to establish a continuity between SAC activity, whole-cell currents and physiological activity measured by affects on the action potential. Agents that blocked SACs (Gd^{3+} and *Grammostola spatulata* venom) also blocked whole-cell currents. Suction on tight seals and loose seals affected action potential generation, and the reversal potentials of the two single-channel SAC types observed in the preparation straddled the whole-cell reversal potential. Short of using more specific blockers, or cloning and knockout of MSCs from this preparation, the physiological role of SACs in this preparation is quite strong.

Although dangerously far from experiment, stretch induced-effects on action potentials in heart cells can be modeled by computer simulations. The results suggest that most of the observed changes can be accounted for by the presence of a population of SACs with a 25 pS conductance, a density of $0.3/\mu m^2$, a reversal potential of about –20 mV, and an operational range of open probability ranging from % to $\approx$30% (Sachs 1994; Zabel et al. 1996).

6.2.3
Unresolved Stories of Vertebrates

Most speculations about the physiological roles for MSCs have yet to be confirmed. For example, a physiological raison d'etre for the mechanosensitivity of the *Xenopus* oocytes SACs has not yet been found. Repeated efforts to determine whether they are required during early embryogenesis have failed to implicate the channels (Steffensen et al. 1991; Wilkinson et al. 1996). Osmotic swelling that activates anion channels in the *Xenopus* oocytes did not produce the cation currents that would be expected if the SACs were also activated (Ackerman et al. 1994). Smooth muscle from various preparations exhibit nonselective, K-selective and Ca-selective mechanosensitive currents (Hisada et al. 1993; Dopico et al. 1994; Ordway et al. 1995; Langton 1993). Frustratingly, the first type are evident at the single-channel level but not have not been reported in whole-cell recordings. Conversely, the mechanosensitive Ca^{2+} currents have only been reported for whole-cell recordings, and these same recordings yielded no evidence of current through nonselective cation channels.

6.2.4
Invertebrates

The crayfish stretch receptor was one of the earliest mechanoreceptors to be studied electrophysiologically and much work has been done on the whole-cell generator currents (Brown et al. 1978; Swerup 1983; Edwards et al. 1981; Rydqvist and Purali 1993). SACs have been recorded from the stretch receptor (Erxleben 1989), and their properties would appear to fit the requirements for the source of the generator current. Unfortunately for completion of the story, the single-channel recordings could not be made from the distal dendrites of the receptor itself, but only from membrane closer to the soma.

C. elegans possesses behavioral withdrawal responses to touch that have been used to isolate a number of touch insensitive mutants mentioned above (see Hong and Driscoll 1994; Huang and Chalfie 1994). Unfortunately, *C. elegans* does not lend itself to electrophysiology, so there are no direct data on MSC function.

Drosophila is known to have MSCs in striated muscle (Zagotta et al. 1988), and mechanosensitive mutants have been constructed that exhibit a block of the behavioral response and the mechanosensitive field potentials of the antennae (Kernan et al. 1994). These mutants, however, appear to maintain SAC activity in the muscles (M. Sokabe, personal communication).

6.2.5
Protozoans

In general, all organisms are sensitive to the effects of gravity, but in most cases the transduction mechanisms are not known (Machemer and Braucker 1992). *Euglena*, the combination plant-animal often studied in introductory zoology courses, is gravitropic. The gravitropism is blocked

by Gd^{3+}, suggesting (to those authors, at least) that MSCs are involved
(Lebert and Hader 1996). Unfortunately, no direct patch clamp records
have been made. *Paramecium* has two sets of mechanosensitive conductances, one Ca^{2+} selective and the other, K^+ selective. The two conductances are expressed in opposing anterior/posterior gradients so that
bumping the anterior causes reversal of the cilia and bumping the poste-
rior causes an increased beating rate, a "getaway" response (Machemer and
Braucker 1992). Again, no single-channel recordings have been made from
Paramecium.

6.2.6
Plants

MSCs are known to be present in higher plants (Falke and Misler 1988;
Ding and Pickard 1993a,b), and it would seem possible that they are
involved in geotropism (Edwards and Pickard 1987), osmotic balance and
elongation at growth points. The most dramatic use may occur in the
touch sensitive plants such as the Venus fly trap and the *Mimosa*. Unfortunately, none of these preparations has been studied at the single-channel
level. Some genes in plant are known to be activated by touch (Braam and
Davis 1990) and these genes seem to be coupled to Ca^{2+} homeostasis. This
is an obvious place for MSCs to play a role, but again, no measurements
have been done.

In the lower plants and fungi there are many interesting behaviors that
involve mechanical transduction. One of the better studied is the rust
fungus that seeks entrance to a plant leaf through the pore like stomata.
The fungus finds the stomata based upon the fact that it is 0.5 μm above
the surrounding leaf (Hoch et al. 1987). These fungi have MSCs that perform much like MscL of *E. coli* (Zhou et al. 1991), but it has not yet been
possible to show a direct correlation between the ridge seeking behavior
and the channels. Even in the case of *E. coli*, for which the proposed
function of the channel was to open in response to extreme osmotic
swelling (Martinac 1993), knockout of MscL by mutagenesis (Sukharev et
al. 1994a) has revealed no marked phenotypic changes. This result, quite
similar to the negative results of many other knockout experiments, may
be explained by the presence of different families of MSCs that replace the
functional activity of the missing MscL.

7
Concluding Remarks

What can we assert with confidence about MSCs? First, we know that most cell types exhibit one or more species of MSCs – channels whose open probability can be substantially modified by suction applied through a patch pipette. In only a small fraction of cases has it been shown that mechanical stimulation of whole cells elicits currents that appear to flow through the MSCs seen in the patches. We must conclude that the physiological significance of most MSC activity remains to be determined.

Second, we know that MSCs are extremely diverse at a molecular level. We can state this even though only one MSC (from *E coli*) has been cloned and expressed. The cloned *E coli* channel shows no sequence similarity to eucaryotic DNA and, in any case, eucaryotic MSCs include nonselective cation channels, K-selective channels and anion-selective channels. In addition, two radically different eucaryotic channels, which have been cloned (NMDA and G-protein regulated K^+ channels) are mechanosensitive when liganded.

Third, we know unequivocally that MSCs can function as physiological mechanotransducers. While this is not a surprise, earlier assertions of this sort were been based on a priori reasoning and partial evidence. What was missing was a continuity of evidence from the single-channel MSC level through to the whole-cell level. Happily, that continuity has now been provided in two cases: the hypothalamic osmoreceptor neurons and chick heart cells. MSC activity underlies the osmosensory currents seen in conventional whole-cell recordings; currents that are capable of affecting vasopressin release and also the ability of stress to alter the action potential properties of heart cells. An explanation for the responsiveness of MSCs in terms of molecular and cytological architecture is now awaited.

Where are we uncertain in important ways? In summary, we are unequipped to predict the conditions under which channels will show mechanosensitivity. A few examples are: In disrupted cells or in artificial lipid bilayers, *E. coli* MSC channels (MscL) exhibit mechanosensitivity and can even show adaptation, but do the channels' in vivo environment enable them to exhibit mechanosensitivity? Likewise, we are ignorant as to whether any eucaryotic MSC channel is mechanosensitive in an artificial lipid bilayer. In two dissimilar eucaryotic channels, mechanosensitivity appears exclusively when channels are liganded by an agonist, but in general, liganded channels do not show mechanosensitivity. Among diverse MSCs, various aspects of mechanosensitivity (dynamics, sensitivity) depend strongly on the mechanical history of the channel-bearing membrane, as if cortical structures determine how mechanical energy is fed to the channels' gating mechanisms. But we do not know what cortical elements are involved.

What is most lacking? The elusive "holy grail" for MSCs is a knowledge of how transduction occurs. The process is distinctive in requiring major alterations in channel dimensions, but a detailed understanding will require knowing the amino acid sequence of the protein(s), and more importantly, the three dimensional structure of the open and closed channels. If our experience from the *E coli* channel is any guide, sequence is not enough. It seems unlikely that a mechanosensitive gating domain will be obvious from sequence data; the S4 story of voltage sensitive channels is not likely to be repeated. In hair cells, understanding mechanotransduction would be impossible in terms of the channel structure alone – it is imperative to incorporate knowledge of the force coupling structures. There is increasing evidence that membrane structures beyond the channel are critical in signal conditioning of MSCs.

Mechanosensitive gating may be exceedingly simple in molecular and energetic terms. This is suggested by the apparent molecular diversity of MSCs and by the widespread ability of MSCs to retain mechanosensitivity when the membrane/cortex has been disrupted. A gating transition in a protein involves movement of a peptide domain. A gating transition occurs when various bonds, biased by the applied forces, absorb sufficient additional energy from the thermal environment. Some channels have evolved so that agonists or electrical fields will bias their gating transitions. Others use mechanical stress. In many cases, it may be important to *reduce* mechanical sensitivity if a channel serves other functions. MSCs presumably are distinctive in that they do couple efficiently to mechanical inputs. The cytoarchitecture may render a channel either sensitive or insensitive to mechanical stimuli. This leads us to predict what is observed in the literature: a continuum of sensitivities.

As the pursuit for a molecular and biophysical understanding of mechanosensitivity gets seriously underway, it will be worth bearing in mind that Nature has put evolutionary effort into creating mechanosensitive channels and methods for modulating them. This modulation, as it can involve large scale integration of many inputs by the cytoarchitecture, may prove to be one tool by which cells prepare for the future.

Acknowledgements. This work was supported by grants to F. Sachs from the United States Army Research Office and the NIH and to C. E. Morris from the Canadian MRC and NSERC. We would like to thank Sergei Sukarev, Glenna Bett, Wade Sigurdson and Charles Bowman for critical comments and Sid Simon and Tony Yeung for advice.

References

Ackerman MJ, Wickman KD, Clapham DE (1994) Hypotonicity activates a native chloride current in Xenopus oocytes. J Gen Physiol 103:153–179

Akay M, Craelius W (1993) Mechanoelectrical feedback in cardiac myocytes from stretch-activated ion channels. IEEE Trans Biomed Eng 40:811–816

Akoev GN, Alekseev NP, Krylov BV (1988) Mechanoreceptors. Springer, Berlin Heidelberg New York

Ashmore JF (1991) The electrophysiology of hair cells. Ann Rev Physiol 53:465–476

Awayda MS, Ismailov II, Berdiev BK, Benos DJ (1995) A cloned renal epithelial Na+ channel protein displays stretch activation in planar lipid bilayers. Am J Physiol 268:C1450–C1459

Bainbridge FA (1915) The influence of venous filling upon the rate of the heart. J Physiol (Lond) 50:65–84

Bargmann CI (1994) Molecular mechanisms of mechanosensation. Cell 78:729–731

Bear CE (1990) A nonselective cation channel in rat liver cells is activated by membrane stretch. Am J Physiol 258:C421–C428

Bear CE, Li CH (1991) Calcium-permeable channels in rat hepatoma cells are activated by extracellular nucleotides. Am J Physiol 261:C1018–C1024

Bedard E, Morris CE (1992) Channels activated by stretch in neurons of a helix snail. Can J Physiol Pharmacol 70:207–213

Benos DJ, Cunningham S, Baker RR, Beason KB, Oh Y, Smith PR (1992) Molecular properties of amiloride sensitive sodium channels. In: Blaustein MP, Habermann E, Reuter H, Schweiger M (eds) Reviews of Physiology Biochemistry and Pharmacology. Springer, Berlin Heidelberg New York, pp 31–114

Berendsen HJC (1996) Bio-molecular dynamics comes of age. Science 271:954–955

Berrier C, Coulombe A, Houssin C (1992) Gadolinium ion inhibits loss of metabolites induced by osmotic shock and large stretch-activated channels in bacteria. Eur J Biochem 206:559–565

Bloom M, Evans E, Mouritsen OG (1991) Physical properties of the fluid lipid-bilayer component of cell membranes: a perspective. Q Rev Biophys 24:293–397

Boitano S, Sanderson MJ Dirksen ER (1994) A role for Ca2+-conducting ion channels in mechanically induced signal transduction of airway epithelial cells. J Cell Sci 107:3037–3044

Bonhivers M, Guihard G, Pattus F, Letellier L (1995) In vivo and in vitro studies of the inhibition of the channel activity of colicins by gadolinium. Eur J Biochem 229:155–163

Bourque CW, Oliet SH, Richard D (1994) Osmoreceptors, osmoreception, and osmoregulation. Front Neuroendocrinol 15:231–274

Bowman CB, Lohr JW (1996) Curvature sensitive mechanosensitive ion channels and smotically evoked movements of the patch membrane. Biophys J 70:A365

Bowman CL, Ding JP, Sachs F, Sokabe M (1992) Mechanotransducing ion channels in astrocytes. Brain Res 584:272–286

Braam J, Davis RW (1990) Rain-, wind-, and touch-induced expression of calmodulin and calmodulin-related genes in arabidopsis. Cell 60:357–364

Brady AJ (1991) Mechanical properties of isolated cardiac myocytes. Physiol Eur Rev 71:413–428

Bregestovski P, Medina I, Goyda E (1992) Regulation of potassium conductance in the cellular membrane at early embryogenesis. J Physiol (Paris) 86:109–115

Brown HM, Ottoson D, Rydqvist B (1978) Crayfish stretch receptor: and investigation with voltage-clamp and ion-selective electrodes. J Physiol (Lond) 284:155–179

Charles AC, Merrill JE, Dirksen ER, Sanderson MJ (1991) Intercellular signaling in glial cells: Calcium waves and oscillations in response to mechanical stimulation and glutamate. Neuron 6:983–992

Chen V, Guber HA, Palant CE (1994) Mechanosensitive single channel calcium currents in rat mesangial cells. Biochem Biophys Res Commun 203:773–779

Chen C-C, Akoplan AN, Sivilotti L, Colquhoun D, Burnstock G, Wood JN (1995) A P2X purinoceptor expressed by a subset of sensory neutrons. Nature 377:428–434

Chen Y, Simasko SM, Niggel J, Sigurdson WJ, Sachs F (1996) Ca2+ uptake in GH3 cells during hypotonic swelling: the sensory role of stretch-activated ion channels. Am J Physiol 270:C1790–C1798

Craelius W (1993) Stretch-activation of rat cardiac myocytes. Exp Physiol 78:411–423

Dai J, Sheetz MP (1995) Mechanical properties of neuronal growth cone membranes studied by tether formation with laser optical tweezers. Biophys J 68:988–996

Davis MJ, Donovitz JA, Hood JD (1992) Stretch-activated single-channel and whole cell currents in vascular smooth muscle cells. Am J Physiol 262:C1083–C1088

Denk W, Webb WW (1992) Foward and reverse transduction at the limit of sensitivity studied by correlating electrical and mechanical fluctuations in frog saccular hair cells. Hearing Res 60:89–102

Denk W, Webb WW, Hudspeth AJ (1989) Mechanical properties of sensory hair bundles are reflected in their Brownian motion measured with a laser differential interferometer. Proc Natl Acad Sci USA 86:5371–5375

Dennerll, Joshi HC, Steel VL, Buxbaum RE, Heidemann SR (1988) Tension and compression in the cytoskeleton of PC-12: neurites. II. Quantitative measurements. J Cell Biol 107:665–674

Diamond SL, Sachs F, Sigurdson WJ (1994) The mechanically induced calcium mobilization in cultured endothelial cells is dependent on actin and phopholipase. Arterioscler Thromb 14:2000–2009

Ding JP, Pickard BG (1993a) Modulation of mechanosensitive calcium-selective cation channels by temperature. Plant J 3:713–720

Ding JP, Pickard BG (1993b) Mechanosensory calcium-selective cation channels in epidermal cells. Plant J 3:83–110

Docherty RJ (1988) Gadolinium selectively blocks a component of calcium current in rodent neuroblastoma glioma hybrid (NG108–15) cells. J Physiol (Lond) 398:33–47

Dopico AM, Kirber MT, Singer, JJ Walsh, JV Jr (1994) Membrane stretch directly activates large conductance Ca2+-activated K+ channels in mesenteric artery smooth muscle cells. Am J Hypertension 7:82–89

Doroshenko P, Neher E (1992) Volume-sensitive chloride conductance in bovine chromaffin cell membrane. J Physiol (Lond) 449:197–218

Driscoll M (1996) Molecular genetics of touch sensation in C. elegans: mechanotransduction and mechano-destruction. Biophys J 70:A1

Duncan RL, Hruska KA (1994) Chronic, intermittent loading alters mechanosensitive channel characteristics in osteoblast-like cells. Am J Physiol 267:F909–916

Duncan RL, Hruska KA, Misler S (1992) Parathyroid hormone activation of stretch-activated cation channels in osteosarcoma cells (UMR-106.01) FEBS Lett 307:219–223

Edwards C, Ottoson D, Rydqvist B, Swerup C (1981) The permeability of the transducer membrane of the crayfish stretch receptor to calcium and other divalent ions. Neurosci 6:1455–1460

Edwards KL, Pickard BG (1987) Detection and transduction of physical stimulii in plants. In Wagner E, Greppin H, Millet B (eds) The cell surface in signal transduction. Springer, Berlin Heidelberg New York, pp 41–66

Elliott JR, Needham D, Dilger JP, Haydon DA (1983) The effects of bilayer thickness and tension on gramicidin single-channel lifetime. Biochim Biophys Acta 735:95–103

Ermakov YA, Averbakh AZ, Lobyshev VI, Sukharev SI (1996) Effects of gadolinium on electrostatic and thermodynamic properties of lipid membranes. Biophys J 70:A96

Erxleben C (1989) Stretch-activated current through single ion channels in the abdominal stretch receptor organ of the crayfish. J Gen Physiol 94:1071–1083

Erxleben CF (1993) Calcium influx through stretch-activated cation channels mediates adaptation by potassium current activation. Neuroreport 4:616–618

Evans E, Needham D (1987) Physical properties of surfactant bilayer membranes: thermal transitions, elasticity, rigidity, cohesion, and colloidal interactions. J Phys Chem 91:4219–4228

Evans E, Yeung A (1994) Hidden dynamics in rapid changes of bilayer shape. Chem Phys Lipids 73:39–56

Falke LC, Misler S (1988) Ion channel acitvity during osmoregulation in clonal neuroblastoma. Biophy J 53:412a

Falke LC, Misler S (1989) Activity of ion channels during volume regulation by clonal N1E115: neuroblastoma cells. Proc Natl Acad Sci USA 86:3919-3923

Fahlke C, Rudel R (1992) Giga-seal formation alters properties of sodium channels of human myoballs. Pflugers Arch 420:248-254

Filipovic D, Sackin H (1992) Stretch- and volume-activated channels in isolated proximal tubule cells. Am J Physiol 262:F857-F870

Franco A Jr, Lansman JB (1990) Calcium entry through stretch-inactivated ion channels in mdx myotubes. Nature 344:670-673

Franco A Jr, Winegar BD, Lansman JB (1991) Open channel block by gadolinium ion of the stretch-inactivated ion channel in mdx myotubes. Biophys J 59:1164-1170

Franco-Obregon A Jr, Lansman JB (1994) Mechanosensitive ion channels in skeletal muscle from normal and dystrophic mice. J Physiol (Lond) 481:299-309

French AS (1986) The role of calcium in the rapid adaptation of an insect mechanoreceptor. J Neurosci 6(8) 2322-2326

French AS (1992) Mechanotransduction. Ann Rev Physiol 54:135-152

Gallez D, Coakley WT (1986) Interfacial instability at cell membranes. Progr Biophys Mol Biol (Lond) 48:155-199

Gannier F, Beranengo JC, Jacquemond V, Garnier D (1993) Measurements of sarcomere dynamics simultaneously with auxotonic force in isolated cardiac cells. IEEE Trans Biomed Eng 40:1226-1232

Gannier F, White E, Lacampagne A, Garnier D, Le Guennec J (1994) Streptomycin reverses a large stretch induced increase in (Ca)i in isolated guinea pig ventricular myocytes. Circ Res 28:1193-1198

Goligorsky MS (1988) Mechanical stimulation induces Ca2+ transients and membrane depolarization in cultured endothelial cells. Effects on Ca2+ in co-perfused smooth muscle cells. FEBS Lett 240No.12:59-64

Grubmuller H, Heymann B, Tavan P (1996) Ligand binding: Molecular mechanics calculation of streptavidin-biotin rupture force. Science 271:997-999

Gruen DW, Wolfe J (1982) Lateral tensions and pressures in membranes and lipid monolayers. Biochim Biophys Acta 688:572-580

Grunder S, Thiemann A, Pusch M, Jentsch TJ (1992) Regions involved in the opening of CIC-2: chloride channels by voltage and cell volume. Nature 360:759-762

Guharay F, Sachs F (1984) Stretch-activated single ion channel currents in tissue-cultured embryonic chick skeletal muscle. J Physiol (Lond) 352:685-701

Gustin MC (1991) Single-channel mechanosensitive currents. Science 253:800

Gustin MC (1992) Mechanosensitive ion channels in yeast. Mechanisms of activation and adaptation. Adv Comp Environ Physiol 10:19-38

Gustin MC, Zhou X, Martinac B, Kung C (1988) A mechanosensitive ion channel in the yeast plasma membrane. Science 242:762-766

Gyorke S, Fill M (1993) Ryanodine receptor adaptation: control mechanism of Ca2+ release in the heart. Science 260:807-809

Hackney CM, Furness DN (1995) Mechanotransduction in vertebrate hair cells: structure and function of the stereociliary bundle. Am J Physiol 268:C1-13

Hagmann J, Dagan D, Burger MM (1992) Release of endosomal content induced by plasma membrane tension: video image intensification and time lapse analysis. Exp Cell Res 198:298–304

Hamill OP, Lane JW, McBride DW (1992) Amiloride: a molecular probe for mechanosensitive ion channels. Trends Pharmacol Sci 13:373–376

Hamill OP, McBride DW (1995) Mechanoreceptive membrane channels. Am Scientist 83:30–37

Hamill OP, McBride DW Jr, (1992) Rapid adaptation of single mechanosensitive channels in Xenopus oocytes. Proc Natl Acad Sci USA 89:7462–7466

Hansen DE, Craig CS, Hondeghem LM (1990a) Stretch-induced arrhythmias in the isolated canine ventricle. Circulation 81:1094–1105

Hansen DE, Craig CS, Hondeghem LM (1990b) Stretch-induced arrhythmias in the isolated canine ventricle. Evidence for the importance of mechanoelectrical feedback. Circulation 81:1094–1105

Hansen DE, Borganelli M, Stacy GP Jr, Taylor LK (1991) Dose-dependent inhibition of stretch-induced arrhythmias by gadolinium in isolated canine ventricles. Evidence for a unique mode of antiarrhythmic action. Circ Res 69:820–831

Hase CC, Le Dain AC, Martinac B (1995) Purification and functional reconstitution of the recombinant large mechanosensitive ion channel (MscL) of Escherichia coli. J Biol Chem 270:18329–18334

Haws CM, Lansman JB (1991) Developmental regulation of mechanosensitive calcium channels in skeletal muscle from normal and mdx mice. Proc R Soc Lond (Biol) 245:173–177

Haws CM, Winegar BD, Lansman JB (1996) Block of single L-type Ca2+ channels in skeletal muscle fibers by aminoglycoside antibiotics. J Gen Physiol 107:421–432

Hille B (1992) Ionic channels of excitable membranes. Sinauer Association, Sunderland

Hisada T, Ordway RW, Kirber MT, Singer JJ, Walsh JV Jr (1991) Hyperpolarization-activated cationic channels in smooth muscle cells are stretch sensitive. Pflugers Arch 417:493–499

Hisada T, Singer JJ, Walsh JV Jr (1993) Aluminofluoride activates hyperpolarization- and stretch-activated cationic channels in single smooth muscle cells. Pflugers Arch 422:397–400

Hoch H, Staples RC, Whitehead B, Comeau J, Wolf ED (1987) Signaling for growth orientation and cell differentiation by surface topography in Uromyces. Science 235:1659–1662

Holley MC, Ashmore JF (1990) A cytoskeletal spring for the control of cell shape in outer hair cells isolated from the guinea pig cochlea. Eur Arch Otorhinol 247:4–7

Hong K, Driscoll M (1994) A transmembrane domain of the putative channel subunit MEC-4 influences mechanotransduction and neurodegeneration in C. elegans (comments). Nature 367:470–473

Howard J, Hudspeth AJ (1988) Compliance of the hair bundle associated with gating of mechanoelectrical transduction channels in the bullfrog's saccular hair cell. Neuron 1:189–199

Hoyer J, Distler A, Haase W, Gogelein H (1994) Ca2+ influx through stretch-activated cation channels activates maxi K+ channels in porcine endocardial endothelium. Proc Natl Acad Sci USA 91:2367–2371

Hu H, Sachs F (1994) Effects of mechanical stimulation on embryonic chick heart cells. Biophys J 66:A170

Hu H, Sachs F (1995) Whole cell mechanosensitive currents in acutely isolated chick heart cells: correlation with mechanosensitive channels. Biophys J 68:A393

Hu H, Sachs F (1996) Single-channel and whole-cell studies of mechanosensitive currents in chick heart. J Memb Biol   154:205–216

Huang M, Chalfie M (1994) Gene interactions affecting mechanosensory transduction in Caenorhabditis elegans. Nature 367:467–470

Huang M, Gu G, Ferguson EL, Chalfie M (1995) A stomatin-like protein necessary for mechanosensation in C. elegans. Nature 378:292–295

Hudspeth AJ (1989) How the ear's works work. Nature 341:397–404

Izu YC, Sachs F (1991) B-D-Xyloside treatment improves patch clamp seal formation. Pflugers Arch 419:218–220

Johansson B, Mellander S (1975) Static and dynamic components in the vascular myogenic response to passive changes in length as revealed by electrical and mechanical recordings from rat portal vein. Circ Res 36:76–83

Jorgensen F, Ohmori H (1988) Amiloride blocks the mechano-electrical transduction channel of hair cells of the chick. J Physiol (Lond) 403:577–588

Kawahara K (1993) Stretch-activated channels in renal tubule. Nippon Rinsho 51:2201–2208

Kernan M, Cowan D, Zucker C (1994) Gentic dissection of mechanosensory transduction: Mechano reception-defective mutations of drosophila. Neuron 12:1195–1206

Kim D (1992) A mechanosensitive K+ channel in heart cells – activation by arachidonic acid. J Gen Physiol 100(6):1021–1040

Kim D, Sladek CD, Aguado-Velasco C, Mathiasen JR (1995) Arachidonic acid activation of a new family of K+ channels in cultured rat neuronal cells. J Physiol 484:643–660

Kim E, Niethammer M, Rothschild A, Jan YN, Sheng M (1995) Clustering of Shaker-type K+ channels by interaction with a family of membrane-associated guanylate kinases. Nature 378:85–88

Kim YK, Dirksen ER, Sanderson MJ (1993) Stretch-activated channels in airway epithelial cells. Am J Physiol 265:C1306–C1318

Kimitsuki T, Ohmori H (1993) Dihydrostreptomycin modifies adaptation and blocks the mechano-electric transducer in chick cochlear hair cells. Brain Res 624:143–150

Kirber MT, Walsh JV, Singer JJ (1988) Stretch-activated ion channels in smooth muscle: a mechanism for the initiation of stretch-induced contraction. Pflugers Arch 412:339–345

Krapivinsky G, Gordon EA, Wickman K, Velimirovic B, Krapivinsky LC, DE (1995) The G-protein-gated atrial K+ channel IKACh is a heteromultimer of two inwardly rectifying K(+)-channel proteins. Nature 374:135–141

Kubalski A, Martinac B, Ling KY, Kung C (1993) Activities of the ion channels in membranes of E. coli lacking the major lipoprotein. J Membr Biol 131:151–160

Lab MJ (1980) Transient depolarisation and action potential alterations following mechanical changes in isolated myocardium. Cardiovasc Res 14:624–637

Lab MJ, Zhou BY, Spencer CI, Horner SM, Seed WA (1994) Effects of gadolinium on length-dependent force in guinea-pig papillary muscle. Exp Physiol 79:249–255

Lacampagne A, Gannier F, Argibay J, Garnier D, Le Guennec JY (1994) The stretch-activated ion channel blocker gadolinium also blocks L-type calcium channels in isolated ventricular myocytes of the guinea-pig. Biochim Biophys Acta 1191:205–208

Lambert S, Bennett V (1993) From anemia to cerebellar dysfunction. A review of the ankyrin gene family. Eur J Biochem 211:1–6

Lane JW, McBride D, Hamill OP (1991) Amiloride block of the mechanosensitive cation channel in Xenopus oocytes. J Physiol (Lond) 441:347–366

Lane JW, McBride DW Jr, Hamill OP (1992) Structure-activity relations of amiloride and its analogues in blocking the mechanosensitive channel in Xenopus oocytes. British J Pharmacol 106:283–286

Lane JW, McBride DW Jr, Hamill OP (1993) Ionic effects on amiloride block of the mechanosensitive channel in Xenopus oocytes. Br J Pharmacol 108:116–119

Langton PD (1993) Calcium channel currents recorded from isolated myocytes of rat basilar artery are stretch sensitive. J Physiol 471:1–11

Lansman JB, Franco Jr A (1991) What does dystrophin do in normal muscle. J Physiol (Lond) 411:409–411

Lebert M, Hader DP (1996) How Euglena tells up from down. Nature 379:590

Lecar H, Morris C (1993) Biophysics of mechanotransduction. In: Rubanyi GM (ed) Mechanoreception by the Vascular Wall. Futura, Mount Kisco, pp 1–11)

Lehtonen JYA, Kinnunen PKJ (1995) Phospholipase A2: as a mechanosensor. Biophys J 68:1888–1894

Levina NN, Lew RR, Heath IB (1994) Cytoskeletal regulation of ion channel distribution in the tip-growing organism Saprolegnia ferax. J Cell Sci 107:127–134

Lewis RS, Ross PE, Cahalan MD (1993) Chloride channels activated by osmotic stress in T lymphocytes. J Gen Physiol 101:801–826

Liu M, Xu J, Tanswell AK, Post M (1994) Inhibition of mechanical strain-induced fetal rat lung cell proliferation by gadolinium, a stretch-activated channel blocker. J Cell Physiol 161:501–507

Machemer H, Braucker R (1992) Gravireception and graviresponses in ciliates. Acta Protozool 31:185–214

Magleby KL, Stevens CF (1972) The effect of voltage on the time course of end-plate currents. J Physiol 223:151–171

Marchenko SM, Sage SO (1996) Mechanosensitive ion channels from endothelium of excised rat aorta. Biophys J 70:A365

Markin VS, Martinac B (1991) Mechanosensitive ion channels as reporters of bilayer expansion. A theoretical model. Biophys J 60:1120–1127

Martinac B (1993) Mechanosensitive ion channels: biophysics and physiology. In Jackson M (ed) Thermodynamics of cell surface receptors. CRC, Boca Raton. pp 327–351

Martinac B, Adler J, Kung C (1990) Mechanosensitive ion channels of E. coli activated by amphipaths. Nature 348:261–263

Marunaka Y, Tohda H, Hagiwara N, Nakahari T (1994) Antidiuretic hormone-responding nonselective cation channel in distal nephron epithelium (A6). Am J Physiol 266:C1513–C1522

Matsumoto H, Baron CB, Coburn RF (1995) Smooth muscle stretch-activated phospholipase C activity. Am J Physiol 268:C458–465

McBride DW Jr, Hamill OP (1992) Pressure-clamp: a method for rapid step perturbation of mechanosensitive channels. Pflugers Arch 421:606–612

McCobb DP, Fowler NL, Featherstone T, Lingle CJ, Saito M, Krause JES (1995) A human calcium-activated potassium channel gene expressed in vascular smooth muscle. Am J Physiol 269:H767–77

Medina I, Bregestovski P (1991) Sensitivity of stretch-activated K+ channels changes during cell-cleavage cycle and may be regulated by cAMP-dependent protein kinase. Proc R Soc Lond (Biol) 245:2–64

Methfessel C, Witzemann V, Takahashi T, Mishina M, Numa S, Sakmann B (1986) Patch clamp measurements on Xenopus laevis oocytes: currents through endogenous channels and implanted acetylcholine receptor and sodium channels. Pflugers Arch 407:577–588

Moody WJ, Bosma MM (1989) A nonselective cation channel activated by membrane deformation in oocytes of the ascidian Boltenia villosa. J Membr Biol 107:179–188

Morris CE (1992) Are stretch-sensitive channels in molluscan cells and elsewhere physiological mechanotransducers? Experientia 48:852–858

Morris CE (1995) Stretch-sensitive ion channels. In: Sperelakis N (ed) Principles of cell physiology and biophysics. Academis, New York, pp 483–489

Morris CE (1996) Stretch channels whether they meant to be or not to be. Biophys J 70:A1

Morris CE, Horn R (1991a) Single-channel mechanosensitive currents. Science 253:801–802

Morris CE, Horn R (1991b) Failure to elicit neuronal macroscopic mechanosensitive currents anticipated by single-channel studies. Science 251:1246–1249

Morris CE, Sigurdson WJ (1989) Stretch-inactivated ion channels coexist with stretch-activated ion channels. Science 243:807–809

Morris CE, Williams B, Sigurdson WJ (1989) Osmotically-induced volume changes in isolated cells of a pond snail. Comparative Biochem Physiol 92:A 479–483

Munger BL, Ide C (1987) The enigma of sensitivity in Pacinian corpuscles: a critical review and hypothesis of mechano-electric transduction. Neurosci Res 5:1–15

Naruse K, Sokabe M (1993) Involvement of stretch-activated ion channels in Ca2+ mobilization to mechanical stretch in endothelial cells. Am J Physiol 264:C1037–C1044

Nazir S (1994) The role of mechnoelectric feedback in the atrium of the isolated Langendorff-perfused guinea-pig heart, and its pharmacological modulation by streptomycin. Thesis, University of Lond

Nazir SA, Dick DJ, Sachs F, Lab MJ (1995) Effects of G. spatulata venom, a novel stretch-activated channel blocker, in a model of stretch-induced ventricular fibrillation in the isolated heart. Circulation 292:I-641, #3076

Niggel J, Hu H, Sigurdson WJ, Bowman C, Sachs F (1996) Grammostola spatulata venom blocks mechanical transduction in GH3 neurons, Xenopus oocytes and chick heart cells. Biophys J 70:A347

Nishizaka T, Miyata H, Yoshikawa H, Ishiwata S, Kinosita K (1995) Unbinding force of a single motor molecule of muscle measured using optical tweezers. Nature 377:251–254

Olesen S-P (1995) Cell membrane patches are supported by proteoglycans. J Membr Biol 144:245–248

Oliet SH, Bourque CW (1993) Mechanosensitive channels transduce osmosensitivity in supraoptic neurons. Nature 364:341–343

Oliet SHR, Bourque CW (1996) Gadolinium uncouples mechanical detection and osmoreceptor potential in supraoptic neurons. Neuron 16:175–181

Opsahl LR, Webb WW (1994a) Transduction of membrane tension by the ion channel alamethicin. Biophys J 66:71–74

Opsahl LR, Webb WW (1994b) Lipid-glass adhesion in giga-sealed patch-clamped membranes. Biophys J 66:75–79

Ordway RW, Petrou S, Kirber MT, Walsh JV Jr, Singer JJ (1995) Stretch activation of a toad smooth muscle K+ channel may be mediated by fatty acids. J Physiol (Lond) 484:331–337

Palmer LG, Frindt G (1996) Gating of Na channels in the rat cortical collecting tubule: effects of voltage and membrane stretch. J Gen Physiol 107:35–46

Palmer RE, Brady AJ, Roos KP (1996) Mechanical measurements from isolated cardiac myocytes using a pipette attachment system. Am J Physiol 270:C697–C704

Paoletti P, Ascher P (1994) Mechanosensitivity of NMDA receptors in cultured mouse central neurons. Neuron 13:645–655

Pender N, McCulloch CA (1991) Quantitation of actin polymerization in two human fibroblast sub-types responding to mechanical stretching. J Cell Sci 100:187–193

Pleusamran A, Kim D (1995) Membrane stretch augments the cardiac muscarinic K+ channel activity. J Membr Biol 148:287–297

Popp R, Hoyer J, Meyer J, Galla HJ, Gogelein H (1992) Stretch-activated non-selective cation channels in the antiluminal membrane of porcine cerebral capillaries. J Physiol (Lond) 454:435–449

Quasthoff S (1994) A mechanosensitive K+ channel with fast-gating kinetics on human axons blocked by gadolinium ions. Neurosci Lett 169:39–42

Rajala GM, Kalbfleisch JH, Kaplan S (1976) Evidence that blood pressure controls heart rate in the chick embryo prior to neural control J Embryol Exp Morphol 36:685–695

Rajala GM, Pinter MJ, Kaplan S (1977) Response of the quiescent heart tube to mechanical stretch in the intact chick embryo. Dev Biol 61:330–337

Ring A (1992) Monitoring the surface tension of lipid membranes by a bubble method. Eur J Physiol 420:264–268

Ring A, Sandblom J (1988) Evaluation of surface tension and ion occupancy effects on gramicidin A channel lifetime. Biophys J 53:541–548

Riquelme G, Jaimovich E, Lingsch C, Behn C (1982) Lipid monolayer expansion by calcium-chlorotetracycline at the air/water interface and, as inferred from cell shape changes, in the human erythrocyte membrane. Biochim Biophys Acta 689:219–229

Rosenberg PA, Finkelstein A (1978) Interaction of ions and water in gramicidin A channels. J Gen Physiol 72:327–340

Rosenmund C, Westbrook GL (1993) Calcium-induced actin depolymerization reduces NMDA channel activity. Neuron 10:805–814

Rothstein A, Mack E (1992) Volume-activated calcium uptake: its role in cell volume regulation of Madin-Darby canine kidney cell. Am J Physiol 262:C339–C447

Rotin D, Bar-Sagi D, O'Brodovich H, Merilainen J Lehto VP, Canessa CM, Rossier BC, Downey GP (1994) An SH3 binding region in the epithelial Na+ channel (alpha rENaC) mediates its localization at the apical membrane. EMBO J 13:4440–4450

Ruknudin A, Song MJ, Sachs F (1991) The ultrastructure of patch-clamped membranes: a study using high voltage electron microscopy. J Cell Biol 112:125–134

Ruknudin A, Sachs F, Bustamante JO (1993) Stretch-activated ion channels in tissue-cultured chick heart. Am J Physiol 264:H960–H972

Rusch A, Kros CJ, Richardson GP (1994) Block by amiloride and its derivatives of mechano-electrical transduction in outer hair cells of mouse cochlear cultures. J Physiol 474:75–86

Rydqvist B, Purali N (1993) Transducer properties of the rapidly adapting stretch receptor neurone in the crayfish (Pacifastacus leniusculus) J Physiol (Lond) 469:193–211

Sachs F (1987) Baroreceptor mechanisms at the cellular level. Fed Proc 46:12–16

Sachs F (1988) Mechanical transduction in biological systems. Crit Rev Biomed Eng 16:141–169

Sachs F (1994) Modeling mechanical-electrical transduction in the heart. In: Mow VC, Guliak F, Tran-Son-Tray R, Hochmuth RM (eds) Cell mechanics cellular engineering. Springer, Berlin Heidelberg New York, pp 308–328

Sachs F, Feng Q (1993) Gated, ion selective channels observed with patch pipettes in the absence of membranes: novel properties of the gigaseal. Biophys J 65:1101–1107

Sachs F, Lecar H (1991) Stochastic models for mechanical transduction (letter). Biophys J 59:1143–1145

Sachs F, Qin F, Palade P (1995) Models of Ca2+ release adaptation. Science 267:2010–2011

Sackin H (1989) A stretch-activated K+ channel sensitive to cell volume. Proc Natl Acad Sci USA 86:1731–1735

Sackin H (1995) Mechanosensitive channels. Ann Rev Physiol 57:333–353

Sadoshima J, Takahashi T, Jahn L, Izumo S (1992) Roles of mechano-sensitive ion channels, cytoskeleton, and contractile activity in stretch-induced immediate-early gene expression and hypertrophy of cardiac myocytes. Proc Natl Acad Sci USA 89:9905–9909

Saimi Y, Martinac B, Delcour AH, Minorsky PV, Gustin MC, Culbertson MR, Adler J Kung C (1993) Patch clamp studies of microbial ion channels. Methods Enzymol 207:681–691

Sakmann B, Neher E (1983) Geometric parameters of pipettes and membrane patches. In: Sakmann B, Neher E (eds) Single-channel recording. Plenum, New York, pp 37–51

Sasaki N, Mitsuiye T, Noma A (1992) Effects of mechanical stretch on membrane currents of single ventricular myocytes of guinea-pig heart. Jpn J Physiol 42:957–970

Schwiebert EM, Mills JW, Stanton BA (1994) Actin-based cytoskeleton regulates a chloride channel and cell volume in a renal cortical collecting duct cell line. J Biol Chem 269:7081–7089

Sharma RV, Chapleau MW, Hajduczok G, Wachtel RE, Waite LJ, Bhalla RC, Abboud FM (1995) Mechanical stimulation increases intracellular calcium concentration in nodose sensory neurons. Neurosci 66:433–441

Sheetz MP, Dai J (1996) Modulation of membrane dynamics and cell motility by membrane tension. Trends Cell Biol 6:85–89

Sigurdson WJ, Morris CE, Brezden BL, Gardner DR (1987) Stretch activation of a K+ channel in molluscan heart cells. J Exp Biol 127:191–209

Sigurdson WJ, Ruknudin A, Sachs F (1992) Calcium imaging of mechanically induced fluxes in tissue-cultured chick heart: role of stretch-activated ion channels. Am J Physiol 262:H1110–H1115

Sigurdson WJ, Sachs F, Diamond SL (1993) Mechanical perturbation of cultured human endothelial cells causes rapid increases of intracellular calcium. Am J Physiol 264:H1745–H1752

Small DL, Morris CE (1994) Delayed activation of single mechanosensitive channels in Lymnaea neurons. Am J Physiol 267:C598–606

Small DL, Morris CE (1995) Pharmacology of stretch-activated K+ channels in Lymnaea neurones. Br J Pharmacol 114:180–186

Small DL, Morris CE, (1995a) Pore properties of Lymnaea neuron SA K+ channels. J Exp Biol 198:1919–1929

Snowdowne KW (1986) The effects of stretch on sarcoplasmic free calcium of frog skeletal muscle at rest. Biochim Biophys Acta 862:441–444

Sokabe M, Hasegawa N, Yamamori K (1993a) Blockers and activators for stretch activated ion channels of chick skeletal muscle. NY Acad Sci 707:417–420

Sokabe M, Nunogaki K, Naruse K, Soga H (1993b) Mechanics of patch clamped and intact cell-membranes in relation to SA channel activation. Jpn J Physiol 43 [Suppl]1:S197–S204

Sokabe M, Nunogaki K, Naruse K, Soga H (1993c) Mechanics of patch clamped and intact cell-membranes in relation to SA channel activation. Jpn J Physiol 43:71–78

Sokabe M, Sachs F, Jing Z (1991) Quantitative video microscopy of patch clamped membranes – stress, strain, capacitance and stretch channel activation. Biophys J 59:722–728

Stacy GP Jr, Jobe RL, Taylor LK, Hansen DE (1992) Stretch-induced depolarizations as a trigger of arrhythmias in isolated canine left ventricles. Am J Physiol 263:H613–H621

Steffensen I, Bates WR, Morris CE (1991) Embryogenesis in the presence of blockers of mechanosensitive ion channels. Dev Growth Differ 5:437–442

Sukharev SI, Martinac B, Arshavsky VY, Kung C (1993) Two types of mechanosensitive channels in the E.Coli envelope: solublization and functional reconstitution. Biophys J 65:177–183

Sukharev S, Blount P, Schroeder M, Kung C (1996) Multimeric structure of bacterial mechanosensitive channel MscL. Biophys J 70:A366

Sukharev SI, Blount P, Martinac B, Blattner FR, Kung C (1994a) A large conductance mechanosensitive channel in E. coli encoded by MscL alone. Nature 368:265–268

Sukharev SI, Blount P, Martinac B, Kung C (1997) Mechanosensitive channels of Escherichia coli: the MscL gene, protein and activities. Ann Rev Physiol 59:633–657

Svoboda K, Schmidt CF, Schnapp BJ, Block SM (1995) Direct observation of kinesin stepping by optical trapping interferometry. Nature 365:721–727

Swerup C (1983) On the ionic mechanisms of adaptation in an isolated mechanoreceptor – an electrophysiological study. Acta Physiol Scand [Suppl] 520:1–43

Takano H, Glantz SA (1995) Gadolinium attenuates the upward shift of the left ventricular diastolic pressure-volume relation during pacing-induced ischemia in dogs. Circulation 91:1575–1587

Taniguchi J, Guggio WB (1989) Membrane stetch: a physiological stimulator of Ca2+-activated K+ channels in thick ascending limb. Am J Physiol 257:F347–F352

Tank DW, Wu ES, Webb WW (1982) Enchanced molecular diffusibility in muscle blebs. J Cell Biol 92:207–219

Tung L, Parikh SS (1993) Cardiac mechanics at the cellular level. J Biomech Eng 113:492–495

Tung L, Zou S (1995) Influence of stretch on excitation threshold of single frog ventricular cells. Exp Physiol 80:221–235

Ubl J, Murer H, Kolb H-A (1988) Hypotonic shock evokes opening of Ca2+-activated K channels in opossum kidney cells. Pflugers Arch 412:551–553

Van Wagoner DR (1993) Mechanosensitive gating of atrial ATP-sensitive potassium channels. Circ Res 72:973–983

Vandorpe DH, Morris CE (1992) Stretch activation of the Aplysia S-channel. J Membr Biol 127:205–214

Vandorpe DH, Small DL, Dabrowski AR, Morris CE (1994) FMRFamide and membrane stretch as activators of the Aplysia S-channel. Biophys J 66:46–58

Verrey F, Groscurth P, Bolliger U (1995) Cytoskeletal disruption in A6 kidney cells: impact on endo/exocytosis and NaCl transport regulation by antidiuretic hormone. J Membr Biol 145:193–204

Wan X, Harris JA, Morris CE (1995) Responses of neurons to extreme osmo-mechanical stress. J Membr Biol 145:21-31

Watson PA (1990) Direct stimulation of adenylate cyclase by mechanical forces in S49 mouse lymphoma cells during hyposmotic swelling. J Biol Chem 265(12):6569-6575

Wellner MC, Isenberg G (1994) Stretch effects on whole-cell currents of guinea-pig urinary bladder myocytes. J Physiol (Lond) 480:439-448

Wellner MC, Isenberg G (1995) cAMP accelerates the decay of stretch-activated inward currents in guinea-pig urinary bladder myocytes. J Physiol (Lond) 482:141-156

White E, Le Guennec JY, Nigretto JM, Gannier F, Argibay JA, Garnier D (1993) The effects of increasing cell length on auxotonic contractions: membrane potential and intracellular calcium transients in single guinea-pig ventricular myocytes. Exp Physiol 78:65-78

Widdicombe JH, Kondo M, Mochizuki H (1991) Regulation of airway mucosal ion transport. Int Arch Allergy Appl Immunol 94:56-61

Wilkinson NC, McBride DW, Hamill OP (1996) Testing the putative role of a mechano-gated channel in testing Xenopus oocyte maturation, fertilization and tadpole development. Biophys J 70:A349

Winegar BD, Haws CM, Lansman JB (1996) Subconductance block of single mechanosensitive ion channels in skeletal muscle fibers by aminoglycoside antibiotics. J Gen Physiol 107:433-443

Wirtz HRW, Dobbs LG (1990) Calcium mobilization and exocytosis after one mechanical stretch of lung epithelial cells. Science 250:1266-1269

Xia SL, Ferrier J (1995) A calcium signal induced by mechanical pertubation of osteoclasts. J Cellular Physiol 167:148-155

Yang XC, Sachs F (1989) Block of stretch-activated ion channels in Xenopus oocytes by gadolinium and calcium ions. Science 243:1068-1071

Yang XC, Sachs F (1990) Characterization of stretch-activated ion channels in Xenopus oocytes. J Physiol (Lond) 431:103-122

Yeung A (1994) Mechanics of intermonolayer coupling in fluid surfactant bilayers. Thesis, University of British Columbia

Zabel M, Koller BS, Sachs F, Franz MR (1996) Stretch-induced changes in the monophasic action potential: importance of the timing of stretch and implications for stretch-activated ion channels. Cardiovascular Res 32:120130

Zagotta WN, Brainard MS, Aldrich RW (1988) Single-channel analysis of four distinct classes of potassium channels in Drosophila muscle. J Neurosci 8(12) 4765-4779

Zhang Y, McBride DW, Hamill OP (1996) On the nature of mechano-gated channel activity in cytoskeleton deficient vesicles shed from Xenopus oocytes. Biophys J 70:A349

Zhou XL, Stumpf MA, Hoch HC, Kung C (1991) A mechanosensitive channel in whole cells and in membrane patches of the fungus Uromyces. Science 253:1415-1417

Editor-in-charge: Professor M.P. Blaustein

Calcium and Neuronal Death

M. Leist[1] and P. Nicotera[1]

[1] Chair of Molecular Toxicology, Faculty of Biology, University of Konstanz, PO Box 5560-X911, D-78457 Konstanz, Germany

1
Introduction

The key role of calcium ions in the function of excitable tissues was appreciated as early as 1882 by Sidney Ringer. Since then, a plethora of cellular functions have been found to require Ca^{2+} signals or simply the maintenance of a set Ca^{2+} concentration. The major requirement for the signalling function of Ca^{2+} is the existence of a concentration gradient between the interstitial fluid and the interior of the cell. Extracellular concentrations of free Ca^{2+} range between 1.3 and 1.8 mM, whereas the free intracellular Ca^{2+} concentration ($[Ca^{2+}]_i$) is around 100 nM. Thus, viable cells maintain a concentration gradient across the plasma membrane of more than four orders of magnitude. Various physiological stimuli increase $[Ca^{2+}]_i$ transiently and thereby induce cellular responses. However, under pathological conditions, changes of $[Ca^{2+}]_i$ are generally more pronounced and sustained. Ca^{2+} overload activates hydrolytic enzymes, leads to exaggerated energy expenditure, impairs energy production, initiates cytoskeletal degradation and ultimately results in cell death. Such Ca^{2+}-induced cytotoxicity may play a major role in several neuropathological phenomena including chronic neurodegenerative diseases, as well as acute neuronal losses, e.g. during the pathogenesis of stroke.

2
Historical Perspective

The idea that Ca^{2+} may be cytotoxic dates back to Fleckenstein's suggestion in 1968 that excessive entry of Ca^{2+} into myocytes may be the underlying mechanism of cardiac pathology following ischemia (for review see Fleckenstein 1984). In 1979, it was shown that agonist overstimulation (Leonard and Salpeter 1979) or cytotoxic xenobiotics may cause lethal Ca^{2+} entry into cells (Schanne et al. 1979). From the early 1980s on, the role of Ca^{2+} in cell death was examined intensively, particularly in isolated hepatocytes, in kidney and in brain (Nicotera et al. 1992, 1994; Trump and Berezesky 1995; Siesjö 1981; Siesjö and Bengtsson 1989). It became evident that cellular Ca^{2+} overload may involve multiple intra- and extracellular routes, most of which are also used for physiological signalling. In addition, Ca^{2+} was found to be sequestered into separate intracellular pools that can contribute to Ca^{2+}-induced toxicity even in the absence of extracellular Ca^{2+}.

Along with the understanding of the role of Ca^{2+} as a physiological regulator, it soon became clear that not only alterations of the normal Ca^{2+} homeostasis, but also changes in Ca^{2+} signalling would have adverse effects. These include alterations in cell growth, differentiation and sensitivity to the activation of natural cell death, apoptosis. In a large number of experimental paradigms it has now been shown, that: (a) Direct sustained elevation of $[Ca^{2+}]_i$, e.g. by exposure of cells to ionophores, causes cell death. (b) A $[Ca^{2+}]_i$ elevation precedes cell death induced by pathophysiological stimuli. (c) Prevention of $[Ca^{2+}]_i$ elevation during such experiments can inhibit cell death. (d) Alterations of Ca^{2+} signalling pathways (e.g. potentiation or inhibition of Ca^{2+} currents) can result in cytotoxicity. Examples of such cases in the nervous system will be given below.

3
Experimental Approaches

The understanding of the role of Ca^{2+} in cell death has benefited greatly from advances in analytical methods to estimate Ca^{2+} concentrations in individual cells (compiled in Nuccitelli 1994). A major step has been the design and synthesis of indicators that change their fluorescent properties upon Ca^{2+} binding (Grynkiewicz et al. 1985). Such indicators can be loaded into cells in the form of their more hydrophobic, non-fluorescent acetoxymethylesters. Subsequently, enzymatic cleavage releases the active compound that remains trapped inside the cell membrane. The use of dyes such as fura-2, which changes its excitation maximum upon Ca^{2+} binding, allows an estimation of the absolute $[Ca^{2+}]_i$ independently from the intracellular indicator concentration. The determination of $[Ca^{2+}]_i$ is facilitated by the so-called ratiometric approach, i.e. excitation at different wavelengths and calculation of $[Ca^{2+}]_i$ from the ratio of the intensities at a defined emission wavelength. This method has been successfully applied to the measurement of $[Ca^{2+}]_i$ in individual cells in suspension, by fluorescence-activated cell sorter (FACS) analysis or in individual adherent cells (e.g. neurons) by quantitative photometric or video-imaging techniques (Malgaroli et al. 1987). In the latter case, even subcellular fluctuations of $[Ca^{2+}]_i$ can be followed. Recently, the development of multiple wavelength confocal laser scanning microscopes and the two-photon excitation technique have enabled the application of fluorescent imaging techniques to estimate $[Ca^{2+}]_i$ at the subcellular level within individual neurons (Hernández-Cruz et al. 1990) even directly within neuronal tissue (Denk et al. 1996; Svoboda et al. 1997).

Other analytical methods that have proven to be extremely useful for a variety of applications are the measurement of Ca^{2+} concentrations with Ca^{2+}-sensitive microelectrodes or the luminescent protein aequorin (Rizzuto et al. 1995), electrophysiological characterization of Ca^{2+} channels, especially by patch clamp techniques, and the use of the radioisotope [45]Ca for the examination of Ca^{2+} fluxes, Ca^{2+} pools and redistribution phenomena.

Furthermore, the development of pharmacological agonists and antagonists has enabled the study of different pathways of Ca^{2+} trafficking and downstream targets known to be affected by Ca^{2+}. Most recently, molecular biological approaches have helped to characterize the proteins involved in intracellular Ca^{2+} binding, in Ca^{2+} transport across the membranes and those mediating Ca^{2+} signalling. For example, Ca^{2+} response elements have been identified and their activation has been investigated in neurons using appropriate reporter genes (Ginty et al. 1993; Hardingham et al. 1997; Xia et al. 1996).

4
Regulation of Ca²⁺ Concentration

Ca^{2+} is compartmentalized into various intracellular pools (Fig. 1). At each of these locally restricted sites Ca^{2+} can be either soluble or it may be bound to proteins as part of its effector function [e.g. calmodulin; James et al. (1995)], or for storage [e.g. to calbindins; Dowd et al. (1992); Mattson et al. (1991)]. It is generally assumed that the alterations of free Ca^{2+} concentrations (often spatially restricted) are more relevant to signalling and cytotoxicity than changes in the absolute amount of Ca^{2+}.

4.1
Calcium Influx

Ca^{2+} influx from the extracellular space following the steep concentration gradient of this ion may easily raise $[Ca^{2+}]_i$ and elicit toxicity. A detailed knowledge of the routes of entry is essential since raised $[Ca^{2+}]_i$ derived

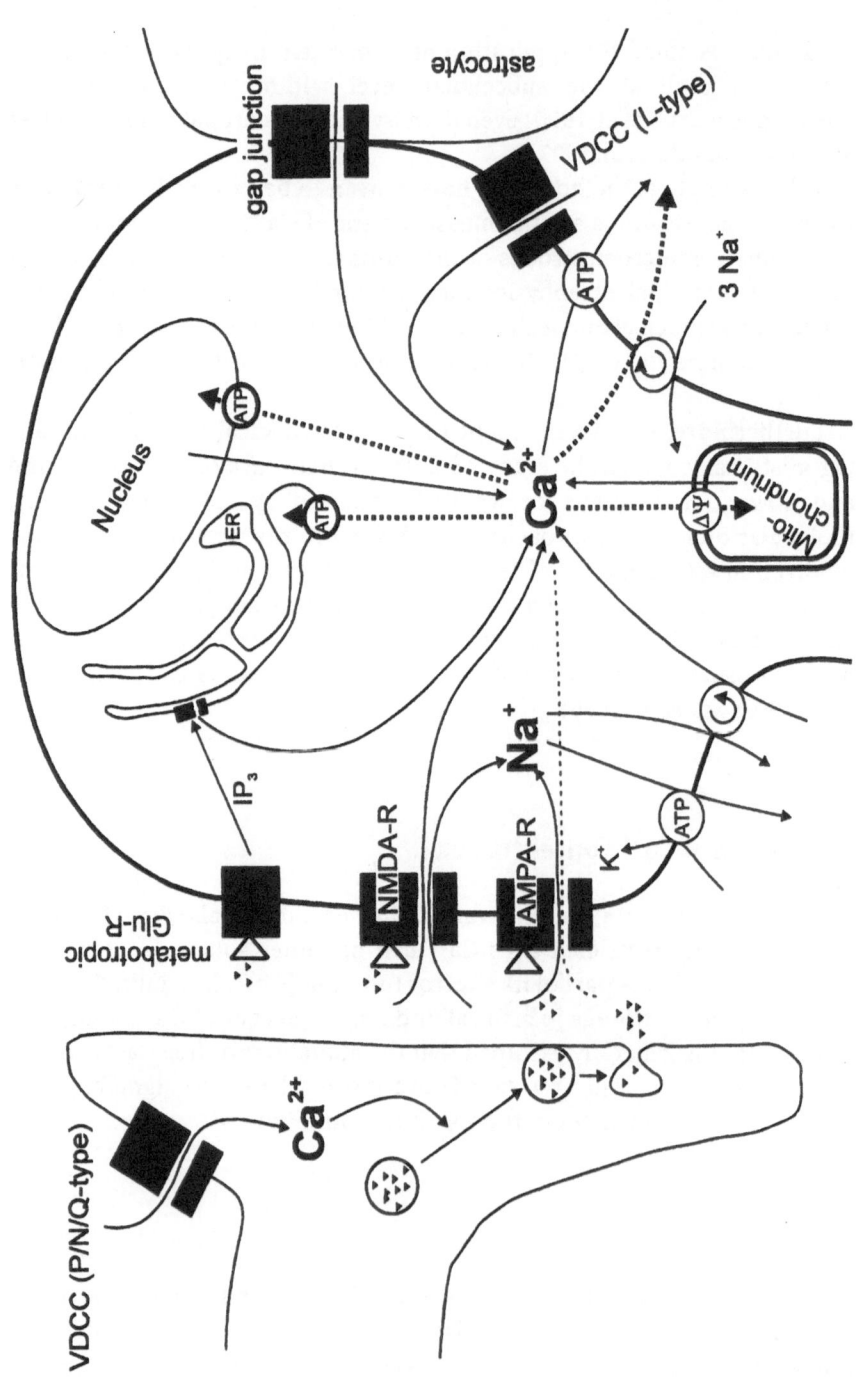

from distinct sources may result in different cellular responses, even with identical average cytosolic concentrations (Ginty et al. 1993; Ghosh et al. 1994; Dugan et al. 1995; White and Reynolds 1995). This may be explained by Ca^{2+} compartmentalization even within the cytosol (Shelanski 1990), by significant and very localized Ca^{2+} gradients near the plasma membrane (Huang and Neher 1996; Tymianski et al. 1994; Borst and Sakmann 1996), by the activation of parallel costimulatory pathways, or by differential effects on cellular Ca^{2+} pools (Hardingham et al. 1997). Also, certain signals involve oscillations of $[Ca^{2+}]_i$ at defined frequencies rather than static changes (Lechleiter et al. 1991).

The main source for Ca^{2+} entry into neurons during activation are voltage- or ligand-gated Ca^{2+} channels. The voltage-dependent Ca^{2+} channels (VDCC) open when the neuronal plasma membrane hypopolarizes. VDCC can be subdivided into several groups (Miljanich and Ramachandran 1995). N-type or P-type channels are involved in presynaptic Ca^{2+} entry necessary for vesicle fusion and neurotransmitter release. They have therefore been implicated in distal neuronal death observed after epileptic attacks (Pollard et al. 1994), in global ischemia (Valentino et al. 1993) or after lesions of dopaminergic neurons by 6-hydroxydopamine or 1-methyl-4-phenylpyridinium (MPP^+) (Mitchell et al. 1994; Cooper et al. 1995; Turski et al. 1991). L-type channels are involved in post-synaptic depolarization and their contribution to neurotoxicity is also well-established (Sucher et al. 1991a; Dreyer et al. 1990; Rossi et al. 1997).

Amongst the ligand-activated Ca^{2+} channels, the N-methyl-D-aspartate (NMDA) receptor (NMDA-R) is probably the most important for neuronal cell death (Rothman and Olney 1995; Choi 1988a, b, 1992, 1995; Choi and Rothman 1990; Hahn et al. 1988; Simon et al. 1984a). The NMDA-R is a heteromeric molecule belonging to the class of ionotropic glutamate receptors (Hollmann and Heinemann 1994). Upon agonist stimulation the NMDA-R channel opens to Na^+ and Ca^{2+}. Interestingly, this ligand-gated

---

Fig. 1. Ca^{2+} fluxes and compartmentalization. Ca^{2+} may enter neurons through voltage-dependent calcium channels (*VDCC*) upon depolarization. Glutamate receptors of the N-methyl-D-aspartate receptor (*NMDA-R*) subtype and of the amino-3-hydroxy-5-methyl-4-isoxazolepropionic acid receptor (*AMPA-R*) subtype constitute the major ligand-gated Ca^{2+}-channels. Intracellular calcium may be sequestered into different organelles by energy-dependent processes. Increase of synaptic calcium concentrations contributes to the triggering of neurotransmitter release from synaptic vesicles. *ER*, endoplasmic reticulum

channel is also voltage-controlled (potential-dependent Mg^{2+} block). The subunit composition and the resulting gating characteristics differ between different brain regions. Kainate or quisqualate receptors, i.e. other classes of ionotropic glutamate receptors involved in excitotoxicity (Pollard et al. 1994; Sheardown et al. 1990), may also function as Ca^{2+} channels (Brorson et al. 1994, 1995; Holzwarth et al. 1994; Geiger et al. 1995). This may be due to special subunit reorganization (Gu et al. 1996; Marin et al. 1993) or to an amino acid exchange reducing the specificity of the channel for Na^+ and allowing the influx of Ca^{2+}. Such amino acid exchange is due to post-transcriptional RNA editing of α-amino-3-hydroxy-5-methyl-4-isoxazolepropionic acid (AMPA)/kainate receptor subunits (Seeburg 1993).

Further possible routes of Ca^{2+} entry are gap junctions between neurons and glial cells (Nedergaard 1994) or the electrogenic Na^+/Ca^{2+} exchanger (Dumuis et al. 1993; Carini et al. 1994) that is activated in conditions of intracellular sodium overload.

4.2
Calcium Sequestration/Export

Cytosolic free Ca^{2+} is sequestered into different organelles largely by energy-consuming processes (for review see Orrenius et al. 1992; Pozzan et al. 1994; Gunter et al. 1994). The uptake systems with the highest affinity are located in the endoplasmic reticulum. At high Ca^{2+} concentrations, potential-driven uptake into mitochondria may also be active. A third independent pool of Ca^{2+} is localized in the cell nucleus (Nicotera et al. 1989, 1990) and its regulation is still controversial. In addition to Ca^{2+} sequestration into organelles, excessive amounts of this cation can also be removed by extrusion via H^+ or Na^+ antiporters (White and Reynolds 1995; Andreeva et al. 1991; Hartley and Choi 1989; Mattson et al. 1989a) and by a high-affinity ATPase Ca^{2+} pump in the plasma membrane (Carafoli 1991). Continuously elevated $[Ca^{2+}]_i$ can therefore increase cellular ATP consumption.

4.3
Intracellular Ca^{2+} Release and Translocation

$[Ca^{2+}]_i$ may rise not only because of influx from the extracellular space, but also due to inhibition of Ca^{2+} efflux or release from intracellular stores

(Nicotera et al. 1992). In neurons, e.g. in excitotoxic injury, it has been demonstrated that a high percentage of the raised intracellular Ca2+ is derived from intracellular pools rather than external sources (Frandsen and Schousboe 1991; Bouchelouche et al. 1989). Accordingly, damage to hippocampal neurons was reduced in an ischemia model by dantrolene, an inhibitor of Ca2+ release from intracellular pools (Wei and Perry 1996). Thus, local intracellular Ca2+ changes may have a role in cytotoxicity independently from Ca2+ influx through the plasma membrane. For example, a Ca2+ pool located in the nuclear envelope can be released by specific agonists (Nicotera et al. 1990). This may regulate the entry of macromolecules through the nuclear pores (Stehno-Bittel et al. 1995) and may contribute to selective changes of the intranuclear Ca2+ concentration (O'Malley 1994; Nicotera et al. 1989; Przywara et al. 1991; Hernández-Cruz et al. 1990).

Ca^{2+} stored in the endoplasmic reticulum (ER) is released by stimulation of two classes of receptors: the IP_3-receptors and the neuronal ryanodine receptors. Therefore, stimulation of cell membrane receptors not linked to Ca^{2+} channels (e.g. metabotropic glutamate receptors) can result in increased $[Ca^{2+}]_i$ due to the generation of second messengers (Dumuis et al. 1993; Murphy and Miller 1989b; Courtney et al. 1990; Verkhratsky and Shmigol 1996) (see Fig. 1). In addition to excessive Ca^{2+} entry through the plasma membrane, the ER pool of Ca^{2+} plays a significant role in xenobiotic-induced toxicity and oxidant cellular injury. This pool is not static, but a steady state is maintained by constant leakage of Ca^{2+} into the cytosol and reimport via an ATP-driven transporter which is very sensitive to oxidant attack. In addition to the ER, a further source of the toxic increase of $[Ca^{2+}]_i$ lies in the mitochondria, which may release this ion upon depolarization, e.g. by NO (Richter et al. 1994).

4.4
Neuronal Set-Point Hypothesis

Increases of $[Ca^{2+}]_i$ above baseline do not necessarily result in cell death; for example, Ca^{2+} influx through VDCC seems to be better tolerated by neurons than NMDA-R-mediated Ca^{2+} increase. There are even situations in which increased cellular Ca^{2+} concentrations foster neuronal survival (Galli et al. 1995) and depletion of Ca^{2+} may induce cell death (Kluck et al. 1994). The best studied neuronal populations in this respect are peripheral sympathetic and sensory neurons. These neurons have intracellular Ca^{2+} concentrations of about 100 nM and are strictly dependent upon nerve growth factor (NGF) for survival, directly after isolation. Over the course of 3 weeks, $[Ca^{2+}]_i$ rises to about 250 nM. Concomitantly, cells loose their requirement for NGF in order to prevent apoptosis. Artificially raising $[Ca^{2+}]_i$ at the beginning of the culture abolishes the NGF requirement of these neurons. These findings suggest that this neuronal population has a developmentally-regulated Ca^{2+} setpoint, and that the $[Ca^{2+}]_i$ determines the dependence on trophic factors and controls neuronal survival (Koike et al. 1989; Johnson and Deckwerth 1993; Franklin et al. 1995; Johnson et al. 1992; Dubinsky 1992).

5
Effector Systems of Raised Ca^{2+}

Unlike excessive $[Na^+]_i$ that may damage cells by osmotic effects, increased $[Ca^{2+}]_i$ probably does not elicit neuronal death by itself. Rather, downstream reactions are activated. Some are directly dependent on the continous presence of Ca^{2+}, others just require a transient $[Ca^{2+}]_i$ increase as a triggering signal. Their nature and contribution to Ca^{2+}-induced cell death have only been partially elucidated.

5.1
Nitric Oxide Synthase

Nitric oxide synthases (NOS) are cytochrome P450-related enzymes that convert arginine to NO and citrulline. Different classes of isoenzymes exist in brain. Two constitutive forms are expressed in neurons (bNOS) or endothelial cells (eNOS), respectively, and are activated by Ca^{2+}/calmodulin following an increase in $[Ca^{2+}]_i$. Other isoforms (iNOS) in microglia or astroglia are inducible by a variety of stimuli, such as cytokines, and function at basal Ca^{2+} concentrations. Activation of bNOS following Ca^{2+} entry through the NMDA-R has been implicated in excitotoxicity to cortical neuronal cultures (Dawson et al. 1991) and in ischemia due to middle cerebral artery occlusion (Huang et al. 1994). The possible terminal cytotoxic mediator may be peroxynitrite formed from $NO^·$ and $O_2^{-·}$. Some neurons, especially cortical neurons expressing high levels of bNOS, seem to be resistant to NO toxicity (Koh and Choi 1988; Koh et al. 1986), but may kill neighbouring neurons because of their Ca^{2+}-induced NO production (Dawson et al. 1993). In cerebellar granule cells (CGCs), elevated $[Ca^{2+}]_i$

causes both NOS activation and cytotoxicity. However, in CGCs, gluta-mate-triggered, Ca^{2+}-mediated cell death is independent of endogenous NO production (Lafon-Cazal et al. 1993). A rather inverted mechanism has been demonstrated in these neurons: Exposure to NO donors leads to stimulation of NMDA-Rs, probably because NO-related species stimulate the release of endogenous agonists (Leist et al. 1997a). This sort of autocrine stimulation eventually causes apoptosis (Bonfoco et al. 1996).

5.2
Hydrolytic Enzymes

Several hydrolytic enzymes are activated by high $[Ca^{2+}]_i$. The main classes include proteases, DNAses and lipases. The following enzymes have been implicated as effectors of Ca^{2+}-elicited toxicity:

- Calpains are Ca^{2+}-activated cysteine proteases (Saido et al. 1994; Wang and Yuen 1997). They have been implicated in toxic cell death in the liver (Orrenius et al. 1989), and in excitotoxic neuronal death in the brain (Brorson et al. 1994; Traystman et al. 1991; Siman and Noszek 1988). In addition, there are nuclear Ca^{2+}-activated proteases that may have a role in the execution phase of apoptosis (Nicotera et al. 1994; Clawson et al. 1992).

- Calcium-dependent DNAses are responsible for DNA degradation that is frequently observed during apoptosis. Despite several attempts to identify and purify endonucleases involved in apoptosis, the nature of the enzyme(s) responsible for the typical oligonucleosomal DNA cleav-age (Hewish and Burgoyne 1973) is still unclear (Bortner et al. 1995).

- Amongst lipases, the Ca^{2+}-dependent phospholipase A_2 (PLA_2) has been implicated in neurotoxicity. Its activation results in the release of arachidonic acid and related polyunsaturated fatty acids, which are further metabolized by lipoxygenases or cyclooxygenases with con-comitant generation of reactive oxygen species (ROS). In addition, PLA_2 activation generates lysophosphatids that alter the membrane struc-ture. This may facilitate Ca^{2+} influx and Ca^{2+} release from internal stores (Traystman et al. 1991). In neurons, there is a close correlation between Ca^{2+} influx through NMDA-Rs and PLA_2 activation (Dumuis et al. 1988, 1993). The release of arachidonic acid following activation of PLA_2 inhibits glutamate uptake into neurons and glial cells and may therefore prolong the excitotoxic action of this amino acid on its receptors (Volterra et al. 1992).

5.3
Xanthine Oxidase

Sustained elevations of $[Ca^{2+}]_i$, caused for example by ischemia, can promote the conversion of xanthine dehydrogenase to xanthine oxidase, i.e.
the enzyme transfers electrons during the catalytic cycle to molecular
oxygen instead of adenin-nicotin dinucleotides. Under low energy conditions, where large parts of ATP are converted to hypoxanthine, this may
result in a massive generation of ROS. The activation of xanthine oxidase
has been implicated in ischemic neuronal death in vivo and in kainate
toxicity to CGCs in vitro (Coyle and Puttfarcken 1993; Dykens et al. 1987).

5.4
Mitochondria

The interaction of mitochondria and Ca^{2+} in toxicity is complex:
* Generally, mitochondria are able to reduce the cytosolic Ca^{2+} overload
  by sequestering large quantities of Ca^{2+} during stimulation of cells.
  However, due to their uptake of Ca^{2+} that would otherwise act as a
  feed-back inhibitor of influx pathways, functioning mitochondria may,
  under certain circumstances, increase the total amount of Ca^{2+} influx
  following an excitotoxic stimulation (Budd and Nicholls 1996a,b). Ca^{2+}
  is sequestered into mitochondria mainly via a Ca^{2+} uniporter or under
  conditions of Na^+ overload via a Ca^{2+}/Na^+ antiporter. The uniport is
  driven by the membrane-potential and has a high capacity, but a relatively low affinity. The lowest level at which brain mitochondria regulate
  $[Ca^{2+}]_i$ is 300 nM in the presence of spermine and may require even
  higher Ca^{2+} concentrations (1 µM) under unfavorable conditions. Thus,
  it has been assumed that Ca^{2+} is only imported into mitochondria
  during conditions of prolonged stimulation and overload. It seems,
  however, that mitochondria contribute to Ca^{2+} regulation also under
  physiological conditions when transient, local high Ca^{2+} concentrations
  are created nearby. Studies in non-neuronal cells have shown that mitochondria can load Ca^{2+} during physiological agonist stimulation and
  may therefore contribute to lower elevated $[Ca^{2+}]_i$ (Rutter et al. 1993).
  Mitochondria have been shown to be pivotal to the reduction of elevated $[Ca^{2+}]_i$ following excitotoxic glutamate stimulation of cells (Kiedrowski and Costa 1995; White and Reynolds 1995), and mitochondrial
  Ca^{2+} deposits were observed in CGCs lethally challenged with NMDA

(Garthwaite and Garthwaite 1986a) or in hippocampal neurons after stroke (Simon et al. 1984b).

- Under conditions of excessive or prolonged Ca^{2+} overload, the sequestration of this ion into mitochondria contributes to the dissipation of their membrane potential ($\Delta\Psi$) and may eventually cause mitochondrial damage. For example, after stimulation of neurons with glutamate or NMDA, mitochondria loose their $\Delta\Psi$ (Ankarcrona et al. 1995). Such a breakdown of $\Delta\Psi$ associated with permeabilization of the inner mitochondrial membrane to ions [also called permeability transition (PT); Bernardi (1996)], has been shown to be due to Ca^{2+} overload under excitotoxic conditions (Schinder et al. 1996; White and Reynolds 1996) and was partially prevented by cyclosporin A, an inhibitor of PT (White and Reynolds 1996; Ankarcrona et al. 1996).

- Various mechanisms have been postulated to explain Ca^{2+} release from mitochondria (Gunter and Pfeiffer 1990):

  - Mitochondrial Ca^{2+} extrusion is an energy-requiring process (33 kJ/mol) linked to H^+ exchange. The net effect is the import of two H^+ in exchange for one Ca^{2+} exported. Ca^{2+} release from mitochondria is stimulated during oxidative stress. Oxidation of nicotinamide adenine dinucleotide phosphate (NADH), with subsequent ADP-monoribosylation of mitochondrial proteins or formaton of cyclic ADP-ribose, has been suggested as a regulatory mechanism. Such enhanced Ca^{2+} extrusion may be the basis of "Ca^{2+} cycling", i.e. continous uptake and release of Ca^{2+} into mitochondria, which leads ultimately to the dissipation of the membrane potential and to mitochondrial failure (Fig. 2). As a consequence of the breakdown of $\Delta\Psi$ or other forms of damage, mitochondria may release Ca^{2+} into the cytosol and generate ROS. For example, under conditions of NMDA-mediated PT, the efflux of Ca^{2+} from mitochondria prologed the increase of $[Ca^{2+}]_i$ and thus contributed to neuronal demise (White and Reynolds 1996). The interaction of raised $[Ca^{2+}]_i$ and ROS may lead to a vicious circle, since stressed mitochondria, possibly with uncoupled respiratory chain, produce ROS as a consequence of NMDA-R stimulation (Dugan et al. 1995; Reynolds and Hastings 1995) and Ca^{2+} overload (Dykens 1994).

  - A mechanism for mitochondrial Ca^{2+} release fundamentally different from the one described above involves the PT of mitochondria (Fig. 2) (Gunter and Pfeiffer 1990), i.e. opening of a pore in the inner mito-

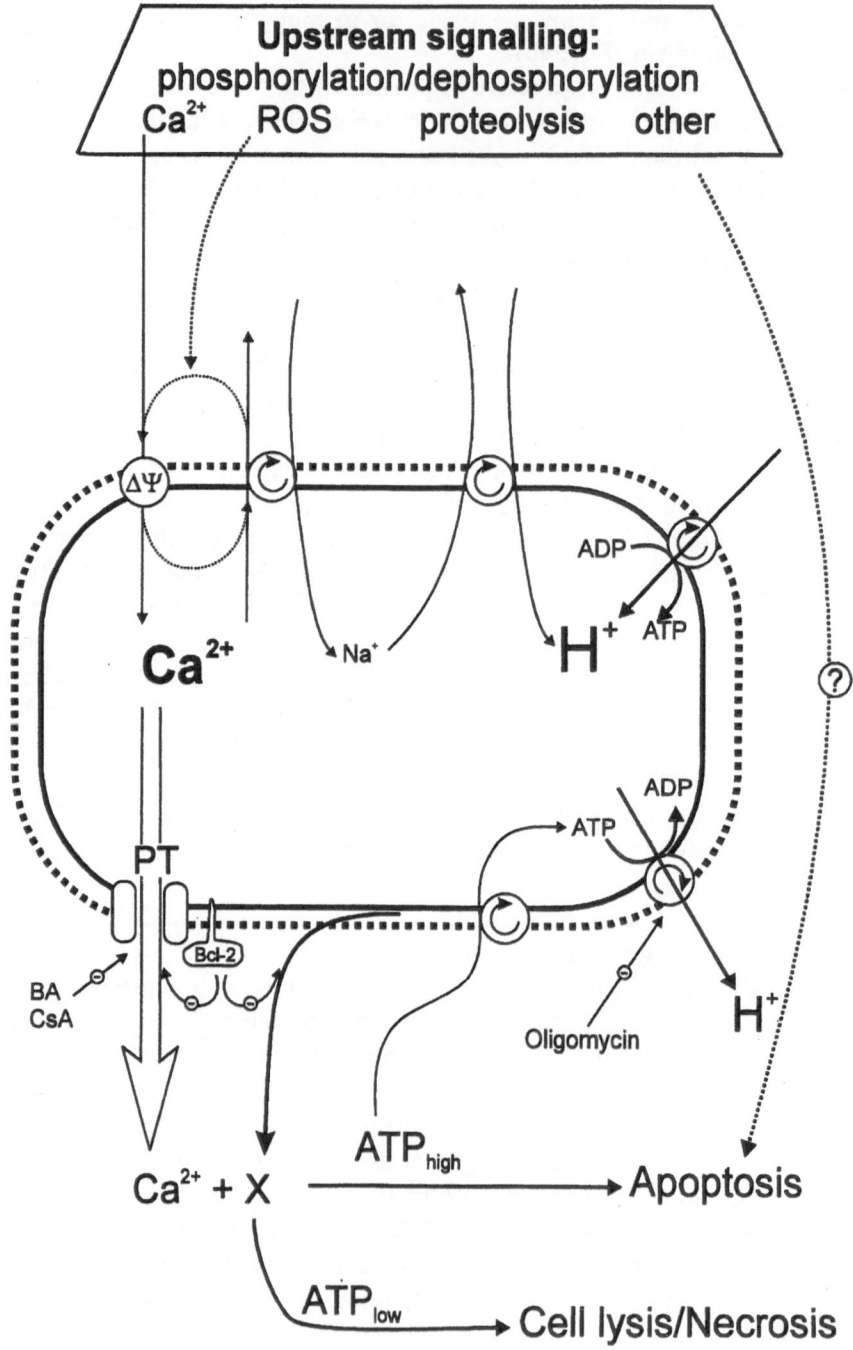

chondrial membrane. Under such conditions mitochondrial Ca^{2+} could be released without energy requirement.

- Furthermore, compromised mitochondria are not only passively involved in cytotoxicity (i.e. because they do not provide the cell with sufficient ATP), but rather they may generate specific signals involved in the execution of apoptosis (see Sect. 6.6). Thus, mitochondria may act as a deciding switch helping cells to recover, accelerating their demise, or actually triggering the execution phase of cell death (Liu et al. 1996; Gunter and Pfeiffer 1990; Zamzami et al. 1996; Newmeyer et al. 1994).

5.5
Protein Phosphorylation/Gene Regulation

The effects of Ca^{2+} on cellular functions and viability do not necessarily depend on the continous elevation of $[Ca^{2+}]_i$. Rather, transient elevation of $[Ca^{2+}]_i$ may trigger persistent effects by altering the state of protein phosphorylation and gene transcription (Ghosh and Greenberg 1995).

At least four different effector systems related to phosphorylation/dephosphorylation may be involved in Ca^{2+}-dependent neurotoxicity: (1) The isoforms type II and type IV of Ca^{2+}/calmodulin-dependent protein kinases (CaMK) are involved in Ca^{2+}-dependent transcriptional regulation. Type II CaMK is also involved in the activation of quisqualate receptors. In addition, this isoenzyme may act as a memory for transient elevations of $[Ca^{2+}]_i$, since its autophosphorylated form remains active even after the return of $[Ca^{2+}]_i$ to basal concentrations. (2) Calcium-sensitive adenylate cyclases, especially of type I, are highly expressed in hippocampus and neocortex and are activated upon glutamate stimulation. (3)

Fig. 2. The mitochondrial role in cell death. Mitochondria act as integrators of various stressful stimuli, such as increased $[Ca^{2+}]_i$ or reactive oxygen species (ROS). Upon stress-induced permeability transition (PT) large anounts of Ca^{2+} plus proteinaceous factors (X = apoptosis-inducing factor and/or cytochrome-c) are released. In isolated mitochondria, PT can be stimulated by atractyloside and prevented by bongkrekic acid (BA) or cyclosporine A (CsA). Possible sites of action for antiapoptotic proteins (Bcl-2) or for agents that inhibit maintenance of mitochondrial membrane potential (oligomycin) using cytosolic ATP are also shown. Although mitochondrial changes seem to be a key switch on the way to cell death, the shape of death (apoptosis or necrosis) seems to be determined by the residual ATP concentrations in the cell (Eguchi et al. 1997; Leist et al. 1997b)

A variety of Ca^{2+}-dependent protein kinase C (PKC) isozymes is expressed in brain. PKC potentiates the NMDA response (Tingley et al. 1993; Cooper et al. 1995), may be involved in delayed Ca^{2+} influx (Manev et al. 1989), and may thereby modify excitotoxic reactions. (4) Calcium-activated phosphatases, e.g. calcineurin, are involved in the regulation of various enzymes (Snyder and Sabatini 1995) that play a role in neurotoxicity. Dephosphorylation of the the NMDA-R by calcineurin limits/shortens Ca^{2+} influx into the cell (Wang and Salter 1994; Tong et al. 1995; Lieberman and Mody 1994). This self-limiting effect on Ca^{2+} entry is blocked by pharmacological intervention with cyclosporine A or FK506. In addition, the extrusion of increased Na^+ following glutamate receptor activation is dependent on the activation of Na^+/K^+-ATPase by calcineurin. Inhibition of Na^+ export with cyclosporine A may therefore preserve ATP (Marcaida et al. 1996). Furthermore, calcineurin dephosphorylates NOS and thereby increases this enzyme activity. A new facet of Ca^{2+} signalling involves the nuclear import of NF-AT transcription factors from their dormant position in the cytoplasm. Ca^{2+}-dependent activation of calcineurin plays a major role in this process (Shibasaki et al. 1996; Timmerman et al. 1996). The overall effect of calcineurin inhibitors on ischemic/excitotoxic neuronal damage seems indeed to be beneficial (Sharkey and Butcher 1994; Dawson et al. 1993; Marcaida et al. 1996; Ankarcrona et al. 1996).

The transcriptional effects of elevated $[Ca^{2+}]_i$ are mediated by various signalling cascades, involving protein kinase A, mitogen-activated protein (MAP) kinases and calmodulin kinase (CaMK). Interestingly, the pathway of Ca^{2+} entry seems to determine the downstream signalling events; for example, brain-derived neurotrophic factor (BDNF) is induced in cortical neurons only by Ca^{2+} entry through VDCC, but not through the NMDA-R (Ghosh et al. 1994). In addition, gene expression is differentially controlled by nuclear and cytosolic Ca^{2+} elevations. This is exemplified (Hardingham et al. 1997) by the induction of the Ca^{2+}-dependent immediate early gene c-*fos* (Morgan and Curran 1986) which has been associated with excitotoxic neuronal death and developmental neuronal apoptosis (Gorman et al. 1995; Smeyne et al. 1993). There are different pathways leading to *fos* activation which may result in different patterns of gene transcription (Ginty et al. 1993; Ghosh and Greenberg 1995; Bading et al. 1993; Ghosh et al. 1994). For example, after stimulation of Ca^{2+} influx through L-type VDCC, nuclear Ca^{2+} concentrations control *fos* expression via the cyclic AMP-responsive element (CRE) which demonstrates that the CRE-binding protein (CREB) can function as a nuclear Ca^{2+}-responsive transcription factor. A second signalling pathway activating *fos* transcription through the serum-responsive element (SRE) is triggered by the rise in cytosolic Ca^{2+} and does not require a large increase in nuclear Ca^{2+} (Hardingham et al. 1997).

5.6
Transglutaminases

Tissue transglutaminase is a Ca^{2+}-dependent enzyme catalyzing the specific crosslinking of proteins during apoptosis (Fesus et al. 1987). Transglutaminase activity has been causally associated with apoptosis of neuroblastoma cells (Piacentini et al. 1992; Melino et al. 1994), and may have an important role in the limitation of focal ischemic brain damage by preventing neuronal secondary lysis and massive release of neurotransmitters (i.e. glutamate).

5.7
Cytoskeletal Components

$[Ca^{2+}]_i$ modifies cytoskeletal organization and dynamics (Mattson et al. 1995). Increased $[Ca^{2+}]_i$ may either directly affect cytoskeletal proteins, or it may change their phosphorylation/polymerization state. In addition, raised $[Ca^{2+}]_i$ may activate proteases cleaving the cytoskeletal elements or proteins necessary for the anchoring of the cytoskeleton to the plasma membrane. Indeed, ischemic hippocampal damage involves proteolysis of fodrin and other cytoskeletal elements and it is reduced by calpain inhibitors (Lee et al. 1991; Siman and Noszek 1988). In cultured cells, high local $[Ca^{2+}]_i$ can cause microtubule depolymerization (Shelanski 1990). In CGCs, NO-induced Ca^{2+} influx triggers the sequential depolymerization and degradation of microfilaments, nuclear lamins and microtubuli (Bonfoco et al. 1996).

A further effect of Ca^{2+} on the cytoskeleton is the alteration of cytoskeletal modulation of membrane receptors and channels. Microfilaments are involved in the desensitization of the NMDA-R after stimulation (Johnson and Byerly 1993), i.e. stabilization of F-actin with phalloidin prevents receptor desensitization (Rosenmund and Westbrook 1993). Conversely, depolymerization of microfilaments with cytochalasin prevents the Ca^{2+} influx induced by Alzheimer's disease-related β-amyloid peptides (Furukawa and Mattson 1995). Modifications of $[Ca^{2+}]_i$ have also been observed after treatment of CGCs with the microtubule depolymerizing agent colchicine (Bonfoco et al. 1995a).

6
The Role of Ca²⁺ in Different Paradigms of Neurotoxicity

6.1
Heavy Metal Toxicity

Several metals can trigger selective neurotoxic processes, at least in part, by disturbing Ca²⁺ homeostasis (Nicotera and Rossi 1993). Metals such as tin or mercury may directly cause an increase of $[Ca^{2+}]_i$ either by displacing Ca²⁺ bound to cellular macromolecules, or because they alter the function of proteins involved in Ca²⁺ import, export or sequestration. This may lead directly to acute or chronic neurodegeneration (Viviani et al. 1995). In addition, metals, at low concentrations, may have very subtle effects, not directly disturbing the Ca²⁺ homeostasis of resting neurons, but altering second messenger systems and channels in a way that modifies the cellular Ca²⁺ responses towards physiological stimuli. This may lead to an increased or decreased Ca²⁺ influx upon stimulation, resulting in altered neurotransmitter release, alterations in cell differentiation and eventually sensitization to apoptosis by otherwise non-toxic stimuli (Rossi et al. 1993, 1997; Viviani et al. 1996).

6.2
HIV Coat Protein gp120

HIV infection and the subsequent AIDS syndrome is frequently associated with CNS defects and dementia (Lipton and Rosenberg 1994; Lipton and Gendelman 1995). One of the pathogenic triggers may be the release of the viral gp120 protein (Brenneman et al. 1988). This protein is known to cause neuronal apoptosis in vivo and in vitro (Müller et al. 1992; Kaiser-Petito and Roberts 1995; Bagetta et al. 1995; Gelbard et al. 1995; Toggas et al. 1994). The present experimental evidence suggests that the neurotoxicity of gp120 is not direct, but involves secondary excitotoxic mechanisms resulting in Ca²⁺ overload (Lipton 1992a). Accordingly, NMDA and gp120 synergistically induce neurotoxicity (Lipton et al. 1991), and blockage of VDCC (Dreyer et al. 1990) or NMDA-R channels by various pharmacological tools (Müller et al. 1992; Lipton 1992b) prevents Ca^{2+} overload and gp120-induced cell death. The terminal excitotoxic mediator acting on the NMDA-R may be produced by macrophages (Lipton 1992c) as a consequence of exposure to gp120 (Giulian et al. 1990).

6.3
Excitotoxic Cell Death

6.3.1
Definition

Excitotoxicity is a phenomenon typically encountered in neurons or myocytes, following a stimulation that exceeds the physiologic range with respect to duration or intensity. Typical excitotoxic stimulators are capsaicin, acetylcholine or – most important in the CNS – glutamate. A large variety of chronic neurodegenerative diseases seem to have an excitotoxic component (Choi 1992; Meldrum and Garthwaite 1990). A causal contribution of excitotoxicity to neuronal damage has also been established in stroke or head trauma (Lipton and Rosenberg 1994; Bullock 1995; Myseros and Bullock 1995), as well as in acute poisoning phenomena, e.g. with CO or MPP^{+} (Turski et al. 1991; Ishimaru et al. 1992). Early observation showed that direct injection of glutamate was selectively neurotoxic in vivo (Olney 1969; Lucas and Newhouse 1957). Generally, excitotoxicity is induced by conditions favouring glutamate accumulation in the extracellular space, and it is enhanced by conditions that inhibit cellular protective mechanisms, e.g. by energy depletion (Novelli et al. 1988). Typical conditions leading to increased extracellular glutamate concentrations (Bullock et al. 1995; Rothman 1984; Sandberg et al. 1986; Drejer et al. 1985; Beneviste et al. 1984) are depolarization of neurons, energy depletion due to hypoglycemia or hypoxia (Cheng and Mattson 1991, 1992; Cheng et al. 1994; Wieloch 1985), or defects in the glutamate re-uptake systems (Volterra et al. 1992; Rothstein et al. 1996). The contribution to neurotoxicity of synaptic activity, i.e. the release of glutamate and stimulation of its receptors on postsynaptic membranes, was deduced from experiments showing that inhibition of neurotransmission by Mg^{2+} or glutamate antagonists protected neurons from hypoxia (Rothman 1983, 1984). More recently the link between glutamate-release and increased $[Ca^{2+}]_i$ has been established within individual neurons; it has been shown that neurotransmitter release, triggered by electrical stimulation of hippocampal neurons with

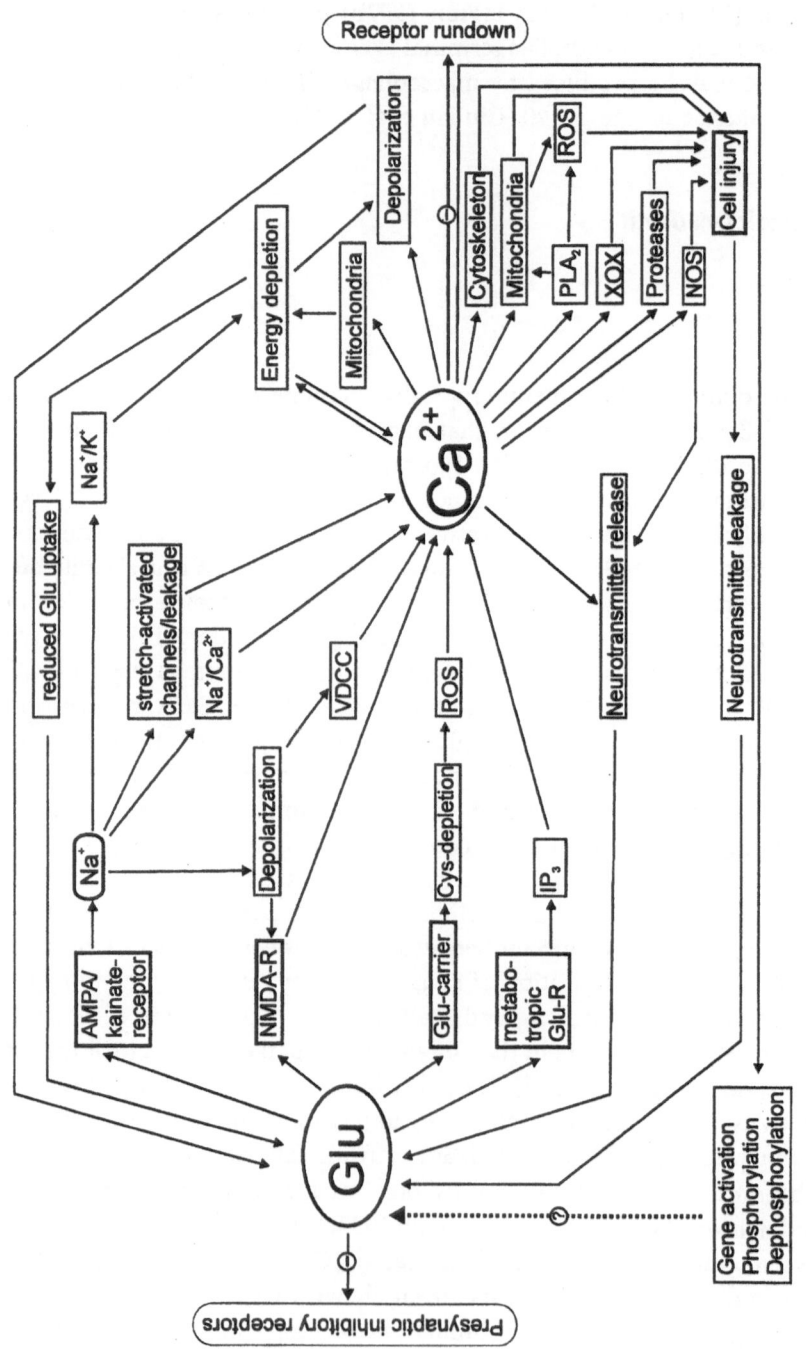

autosynapses, was sufficient to cause glutamate-induced increases of $[Ca^{2+}]_i$ within the same cell (Tong et al. 1995). This close reciprocal relationship between increased extracellular glutamate concentrations and raised $[Ca^{2+}]_i$ may result in neuronal death under excitotoxic conditions (Fig. 3).

6.3.2
Glutamate and Ca^{2+} Influx

Glutamate can trigger Ca^{2+} influx by various routes. A Ca^{2+} channel is directly opened by stimulation of the NMDA-R (Seeburg 1993). The key role of this channel in excitotoxicity is supported by pharmacological intervention studies (Chen et al. 1992; Simon et al. 1984a; Lipton and Rosenberg 1994; Dubinsky 1992; Gill et al. 1992; Michaels and Rothman 1990), and by the protective effect of NMDA-R anti-sense RNA in an experimental stroke model (Wahlestedt et al. 1993). NMDA-R antagonists achieve only a partial protection, especially in models of global ischemia as compared to focal ischemia (Valentino et al. 1993; Myseros and Bullock 1995; Buchan et al. 1991; Buchan and Pulsinelli 1990). Additional mechanisms seem, therefore, to be involved especially in neuronal death in the ischemic core region.

Non-NMDA, ionotropic glutamate receptor subtypes (non-NMDA-R) are other important mediators of excitotoxicity (Brorson et al. 1994, 1995; Holzwarth et al. 1994). Non-NMDA-R agonists cause Ca^{2+} influx into neurons (Gu et al. 1996; Hajos et al. 1986b; Murphy and Miller 1989a,b; Courtney et al. 1990; Pastuszko and Wilson 1985) and neurotoxicity in vivo as well as in vitro (Dykens et al. 1987; Manev et al. 1989; Frandsen et al. 1989; Cox et al. 1990; Milani et al. 1991; Marcaida et al. 1995; Sucher et al. 1991b; Bindokas and Miller 1995). The mechanisms may involve direct opening of

---

Fig. 3. The interactions of $[Ca^{2+}]_i$ and glutamate receptor agonists in neuronal injury. Release of glutamate causes an increase of $[Ca^{2+}]_i$ in postsynaptic neurons by various direct and indirect mechanisms. Conversely, increased $[Ca^{2+}]_i$ causes various cellular changes leading to a release of glutamate. In addition, increased $[Ca^{2+}]_i$ can lead to the activation of various cytotoxic processes that may result in cell death. In pathologic situations these relations may constitute a vicious circle forming the basis of tissue destruction. *NOS*, nitric oxide synthase; *XOX*, xanthine oxidase; *PLA2*, phospholipase A2; *IP3*, inositol trisphosphate; Na^+/Ca^{2+}, Na^+/Ca^{2+} exchanger; Na^+/K^+, Na^+/H^+ ATPase; *ROS*, reactive oxygen species; *VDCC*, voltage-dependent Ca^{2+} channels

Ca^{2+} channels (Mattson et al. 1989b, 1993; Gu et al. 1996), the release of glutamate, which subsequently acts on NMDA-Rs (Pollard et al. 1994; Courtney et al. 1990; Rothman 1984; Sandberg et al. 1986; Drejer et al. 1985; Beneviste et al. 1984; Levi et al. 1991; Chittajallu et al. 1996; Sucher et al. 1991b; Mattson et al. 1993a) or a Ca^{2+} release from intracellular stores (Murphy and Miller 1989b). An additional mechanism operated by non-NMDA-R may be an excessive Na^+ influx (up to 60 mM intracellular concentration), which subsequently causes destabilization of Ca^{2+} homeostasis (Kiedrowski et al. 1994; Bindokas and Miller 1995; Reuter and Porzig 1995). In addition, depolarization of neurons following non-NMDA-R stimulation releases the Mg^{2+} block of the NMDA-R (Monyer et al. 1992a). Ca^{2+} permeable AMPA/kainate receptors are also permeable to Co^{2+} ions. This property can be used to visualize Ca^{2+} influx through non-NMDA-R directly within individual neurons (Lu et al. 1996).

Besides the non NMDA-R, further ligand-operated ion channels may directly control excitotoxic Ca^{2+} influx. Important examples are the aspartate receptor on Purkinje cells (Yuzaki et al. 1996) and capsaicin-stimulated receptors on sensory neurons (Chard et al. 1995).

Further mechanisms may contribute indirectly to raise $[Ca^{2+}]_i$. Glutamate may, for example, increase oxidative stress and trigger subsequent death in neurons by inhibition of cystine uptake and subsequent thiol depletion (Ratan et al. 1994). Conversely, increased extracellular cystine concentrations may cause a glutamate release by reversal of the uptake system and subsequent excitotoxic stimulation of neurons (Piani and Fontana 1994). In addition, neuronal depolarization following stimulation of glutamate receptors may activate Ca^{2+} influx through VDCCs (Sucher et al. 1991a; Dreyer et al. 1990; Bührle and Sonnhof 1983).

Besides the ionotropic glutamate receptors that have an established role in the deregulation of Ca^{2+} homeostasis and excitotoxicity, there is also a large class of metabotropic glutamate receptors that control intracellular Ca^{2+} concentration. Depending on the subtype of receptor, stimulation may be neuroprotective (Pizzi et al. 1996; Maiese et al. 1996) or aggravating neurotoxicity (Mukhin et al. 1996).

6.3.3
Excitotoxicity and Raised $[Ca^{2+}]_i$

The key role of Ca^{2+} in excitotoxicity (Siesjö and Bengtsson 1989; Dubinsky 1992; Choi 1995; Gibbons et al. 1993) is suggested by three different lines of evidence: (1) There is an obvious increase in $[Ca^{2+}]_i$ in in vivo and in in vitro models of excitotoxic cell death. This has been observed in ischemic brain (Simon et al. 1984b; Dienel 1984) or in brain slices exposed to NMDA-R agonists or anoxia (Kass and Lipton 1986; Garthwaite and Garthwaite 1986b). In addition, glutamate-stimulated Ca^{2+} influx has been shown directly in cultured neurons by the [45]Ca technique (Eimerl and Schramm 1994; Berdichevsky et al. 1983; Mogensen et al. 1994; Wroblewski et al. 1985), and increased $[Ca^{2+}]_i$ after NMDA-R stimulation has been observed repeatedly using fluorescent probes (Milani et al. 1991; Tymianski et al. 1993b; De Erausquin et al. 1990; Dubinsky and Rothman 1991; Murphy et al. 1987; Michaels and Rothman 1990; Dubinsky 1993). Additional evidence comes from in vivo microelectrode measurements, demonstrating an 80%–90% decline in extracellular Ca^{2+} during ischemia and a corresponding increase in intracellular Ca^{2+} (Gill et al. 1992; Kristián et al. 1994; Heinemann and Pumain 1980; Marciani et al. 1982; Bührle and Sonnhof 1983). Although radioactive and fluorescent Ca^{2+} measurements yield quantitatively similar results, it is notable that the absolute amount of Ca^{2+} taken up by cells reaches several millimolar, whereas only changes in the nanomolar range are observed when $[Ca^{2+}]_i$ is monitored (Eimerl and Schramm 1994; Lu et al. 1996). (2) Prevention of Ca^{2+} entry into the cell by removal of extracellular Ca^{2+} (Hartley and Choi 1989; Manev et al. 1989; Hajos et al. 1986b; Garthwaite and Garthwaite 1986b; Rothman et al. 1987; Choi 1985, 1987), depletion of NMDA-Rs (Wahlestedt et al. 1993), or by pharmacological inhibition of glutamate receptors or VDCCs (Choi 1988a, 1995; Lipton and Rosenberg 1994) prevents neuronal death in many paradigms of excitotoxicity. (3) Prevention of neurotoxicity by inhibition of downstream effects of Ca^{2+} overload strongly suggests a causal role of Ca^{2+} in excitotoxicity. Intracellular Ca^{2+} chelators can prevent ischemic damage in vivo and excitotoxic neuronal damage in vitro (Tymianski et al. 1993b, 1994). Also, inhibition of effectors of Ca^{2+} toxicity such as calmodulin (Marcaida et al. 1995), calcineurin (Marcaida et al. 1996; Dawson et al. 1993), or bNOS (Dawson et al. 1993) protects neurons from the toxicity of excitatory amino acids.

6.3.4
Open Issues

Some issues concerning the role of Ca^{2+} in excitotoxicity remain to be resolved. Results in different experimental models are often inconsistent. For example, only some studies show a quantitative correlation between increased $[Ca^{2+}]_i$ and excitotoxicity (Milani et al. 1991; Randall and Thayer 1992; Frandsen et al. 1989; Lu et al. 1996), whereas others fail to do so in similar culture systems (for review see Dubinsky 1992). It is apparent that minor alterations in cell culture conditions seem to have a major influence on the final outcome of experiments. The following experimental problems are frequently encountered:

- The measurement of average $[Ca^{2+}]_i$ neglects the fact that pronounced Ca^{2+} gradients exist within the cell (Connor et al. 1988) and that Ca^{2+} concentrations elicited by glutamate within the dendritic spines stimulated by glutamate may by far exceed the average cytoplasmic concentrations. In fact, localized Ca^{2+} influx, followed by localized blebbing of the plasma membrane, has been observed in cultured neurons stimulated with kainate (Bindokas and Miller 1995) and within single synapses (Reuter 1995). In addition, it is likely that the targets of Ca^{2+} relevant for the induction of cytotoxicity are not homogeneously distributed. This is suggested by different protective effects of fast and slow-acting Ca^{2+} chelators (Tymianski et al. 1994).
- The source of Ca^{2+} entry may determine the cellular fate (Tymianski et al. 1993a; Dugan et al. 1995; White and Reynolds 1995).
- Changes of pH (Kristián et al. 1994), protein expression (Mattson et al. 1991), as well as the presence of astrocytes (Vilbulsreth et al. 1987) and many intercellular mediators have been shown to prevent alterations of neuronal Ca^{2+} homeostasis (Mattson et al. 1993a-c; Cheng and Mattson 1991, 1992; Cheng et al. 1994).
- Kinetics of Ca^{2+} elevation may be very complex and measuring $[Ca^{2+}]_i$ at just one defined time point may not yield sufficient information. It is important to consider that Ca^{2+} may modify its own homeostasis by receptor activation/inactivation (Tong et al. 1995; Ehlers et al. 1996) by causing the release of excitotoxic neurotransmitters or by causing further Ca^{2+} release from intracellular pools. The understanding of these phenomena, especially in the in vivo situation, is still a major research task.

6.3.5
A Vicious Circle Formed by Increased $[Ca^{2+}]_i$ and Release of Glutamate

On the one hand, there are various pathways resulting in increased $[Ca^{2+}]_i$ as a consequence of raised extracellular glutamate concentration. On the other hand, raised Ca^{2+} concentrations and conditions associated with raised $[Ca^{2+}]_i$ often result in increased extracellular glutamate concentrations. This reciprocal relationship between $[Ca^{2+}]_i$ and glutamate creates a vicious circle that may form the basis for neuronal damage due to many different initial insults (see Fig. 3). Accordingly, endogenous glutamate release has been described in many pathological conditions, such as ischemia and after stimulation of neurons with excitotoxins or nitric oxide, or after depolarization (Valentino et al. 1993; Courtney et al. 1990; Wieloch 1985; Choi 1988b; Levi et al. 1984; Gallo et al. 1982; Benveniste et al. 1984; Sandberg et al. 1986; Rego et al. 1996; Drejer et al. 1985; Rothman 1984; Hartley and Choi 1989; Palaiologos et al. 1989; Levi et al. 1991). The control of glutamate release is complex and it is in part inhibited by its own actions on presynaptic receptors (Chittajallu et al. 1996). There are two principally different modes of release. One follows the general mechanisms of exocytosis (Iadecola 1997) and involves fusion of synaptic vesicles with the plasma membrane (Benfenati and Valtorta 1995). The other involves reversal of the glutamate uptake carrier (Nicholls and Attwell 1990). The former is ATP-dependent and can be inhibited by neurotoxins specifically cleaving proteins of the exocytosis machinery (Schiavo et al. 1993, 1995). In line with this, prevention of neurosecretion with tetanus toxin (Monyer et al. 1992b) or botulinum neurotoxin C (Leist et al. 1997a) has proven to be neuroprotective against hypoglycemia or NO challenge.

Exocytosis may be modulated by increased $[Ca^{2+}]_i$ directly or indirectly:

- Synaptic vesicle fusion requires a high local calcium concentration which may be provided by physical association of N-type VDCC with exocytosis proteins (Sheng et al. 1996). Inhibition of presynaptic N-type VDCC reduced ischemic brain damage, possibly by inhibiting glutamate release (Valentino et al. 1993). Modulation of exocytosis by Ca^{2+} is further shown by the role of the Na^+/Ca^{2+} exchanger in synaptic boutons for exocytosis (Reuter and Porzig 1995). The increase of postsynaptic Ca^{2+} following presynaptic depolarization (associated with presynaptic Ca^{2+} influx through VDCC and neurotransmitter release) has been shown on single synapses (Reuter 1995). Moreover, neurotransmitter release following depolarization with K^+ or veratridine has been shown to be Ca^{2+}-dependent and inhibitable by tetanus toxin (Van Vlient et al. 1989).

- Besides its direct effects, increased $[Ca^{2+}]_i$ may cause neurotransmitter release indirectly. For example, nitric oxide synthase, activated by elevated $[Ca^{2+}]_i$, may produce sufficient NO to cause Ca^{2+}-independent neurotransmitter release from synaptic vesicles (Meffert et al. 1994, 1996). This mechanism may explain the increased spontaneous presynaptic glutamate release following stimulation of postsynaptic NMDA-R (Malgaroli and Tsien 1992). Conditions contributing to and aggravating glutamate release are ischemia, mitochondrial damage and depolarization due to energy-depletion (Nicholls and Attwell 1990; Schulz et al. 1996; Rego et al. 1996; Vilbulsreth et al. 1987). Here elevated $[Ca^{2+}]_i$ may contribute to the generation of cellular stress, but not directly to the neurotransmitter release. Under such stress conditions, however, decreased intracellular K^+ reverses the glutamate transporter. This effect is enhanced by increased intracellular Na^+, which blocks the re-uptake of glutamate via the transporter (Nicholls and Attwell 1990).

6.4
Alzheimer's Disease-Associated Amyloid-β-Peptide

Alzheimer's disease is a neurodegenerative disorder with poorly defined mechanisms of pathogenesis. The actions of amyloid-β (Aβ)-peptides derived form amyloid precursor proteins (APP) (Mattson et al. 1993d) have been implicated as a cause of neuropathological changes. The neurodegeneration observed in patients shows apoptotic features (Smale et al. 1995; Lassmann et al. 1995). Neuronal apoptosis is also evoked in vitro by exposure to Aβ-peptides (Forloni et al. 1993; Pike et al. 1993; Le et al. 1995). The mechanism of Aβ-peptide toxicity in vitro may involve excessive Ca^{2+} entry into cells (Mattson et al. 1993e). Furthermore, antigenic changes of cytoskeletal elements typical of Alzheimer's disease are mimicked by excessive Ca^{2+} influx into cultured hippocampal neurons (Mattson 1990). In addition, when neurons were treated by actin depolymerizing agents, which are known to block receptor-mediated Ca^{2+} entry, Aβ-peptide-induced toxicity was prevented (Furukawa and Mattson 1995). The Aβ-induced destabilization of Ca^{2+} homeostasis in hippocampal neurons (Mattson et al. 1992) is corrected by preincubation of neurons with various peptide mediators such as fibroblast growth factor (Mattson et al. 1993b,f) or tumor necrosis factor (Barger et al. 1995). The toxicity of Aβ-peptide is strongly enhanced by conditions also found in ischemic brain injury, i.e. raised concentrations of excitotoxic amino acids and hypoglycemia. It may be speculated that the pathology of Alzheimer's disease is related to a very protracted chronic form of excitotoxicity (Scorziello et al. 1996; Mattson et al. 1993d,e) caused by a conversion of APP to Aβ. Accordingly, the unprocessed APP protein seems to have a dampening effect on the cellular Ca^{2+} by hyperpolarizing neurons via the opening of potassium channels (Furukawa et al. 1996). In agreement with these findings, soluble APP protects neurons from Aβ toxicity (Goodman and Mattson 1994; Mattson et al. 1993g).

6.5
Prion Diseases

The ethiopathology of a variety of spongiform encephalopathies is linked to the change in tertiary structure of an endogenous protein coded by the *prn* gene. Such converted proteins (Prp-Sc) may be infections by themselves and are therefore called prions (Prusiner 1996). Prion infection in vivo or in vitro causes neuronal apoptosis (Forloni et al. 1993; Giese et al. 1995). Similar to the Aβ-peptide, prion protein fragments induce the generation of ROS in mixed neuronal cultures (Brown et al. 1996), and Prp-Sc may cause calcium influx into cortical cells via the NMDA-R (Müller et al. 1993). Although it has been suggested that untransformed prion protein (Prp-c) is necessary for normal synaptic function (Collinge et al. 1994), the causal involvement of chronic calcium overload in diseases due to transformation of Prp-c to Prp-Sc is merely speculative.

6.6
Decision Point: Neuronal Apoptosis vs. Necrosis

The mode of cell death has a large bearing on the fate of the tissue. Apoptosis and necrosis, in their classical definitions, are two fundamentally different modes of cell death (Wyllie et al. 1980). Whereas apoptosis is characterized by a preservation of membrane integrity until the cell is phagocytosed, this is not the case in necrosis/lysis of cells. Therefore, neurotransmitter release and inflammation may ensue in the neighbouring tissue. The duration and extent of Ca^{2+} influx may determine whether neurons survive, die by apoptosis or undergo necrotic lysis (Choi 1995).

Very low $[Ca^{2+}]_i$, or the prolonged inhibition of Ca^{2+} influx, may be neurotoxic (Johnson et al. 1992; Mattson et al. 1995). A continuous moderate increase in $[Ca^{2+}]_i$ as that produced by a sustained slow influx may cause apoptosis, whereas an exceedingly high influx would cause rapid cell lysis. For example, stimulation of cortical neurons with high concentrations of NMDA results in necrosis, whereas exposure to low concentrations causes apoptosis (Bonfoco et al. 1995b). Accordingly, neuronal death in experimental stroke models is necrotic in the ischemic core, but delayed and apoptotic in the less severely compromised penumbra or border regions (Li et al. 1995a–d; Pollard et al. 1994; Beilharz et al. 1995; Charriaut-Marlangue et al. 1996; Linnik et al. 1995). The same applies to several other neuropathologic conditions where apoptosis and necrosis have been observed to occur simultaneously (Shimizu et al. 1996; Gschwind and Huber 1995; Portera-Cailliau et al. 1995; Hartley et al. 1994). The sensor and switch of neurons towards one or the other fate may be located in the mitochondria (Ankarcrona et al. 1995) (see Fig. 2). Ca^{2+} overload or other forms of cellular stress may elicit mitochondrial PT (Gunter and Pfeiffer 1990; Gunter et al. 1994) and a consequent release of Ca^{2+} *plus* a proteinaceous factor related to cell death (Newmeyer et al. 1994; Zamzami et al. 1996; Kluck et al. 1997; Yang et al. 1997; Susin et al. 1996). In this context it is important to note that energization of mitochondria and maintenance of their membrane potential does not necessarily require a functional respiratory chain. ATP may be imported from the cytosol via the ATP/ADP translocator and then generate a membrane potential through the oligomycin-sensitive proton pump (see Fig. 2). Consequently, also mitochondria unable to perform oxidative phosphorylation due to the lack of proteins coded by mitochondrial DNA are able to undergo PT and to induce nuclear apoptotic changes (Zamzami et al. 1996). A complete de-energization of the cell (e.g. failure of all mitochondria and of glycolysis) may not allow the ordered sequence of changes required for the apoptotic demise. In such a case random processes would result in rapid uncontrolled cell lysis/necrosis. Therefore, it seems likely that apoptosis ensues under conditions of Ca^{2+} overload or other stress, where there remains sufficient energy production (ATP) to execute an internal "death programme" (Chou et al. 1995; Hartley et al. 1994). A common finding in apoptosis is, for example, that of morphologically intact mitochondria (Hajos et al. 1986a; Wyllie et al. 1980), that may be energized by electron transport or by import of cytoplasmic ATP. Accordingly, ATP levels are maintained in PC12 cells, in CGC or in hippocampal neurons undergoing apoptosis (Rothman et al. 1987; Ankarcrona et al. 1995; Mills et al. 1995). In a model system using non-neuronal cells, it has been demonstrated recently that ATP is indeed a switch between different shapes of cell death (Leist et al. 1997b).

7
Conclusions

Calcium is an ubiquitous intracellular messenger in neurons. Therefore, it is not surprising that alterations of Ca^{2+} homeostasis are involved in many instances of cell death. In contrast to the role of Ca^{2+} as harbinger of cell death, its role as the executor is still discussed. In various instances, a rise of the $[Ca^{2+}]_i$ may only parallel or follow cell death, without being causally involved. For example, Ca^{2+} influx may be accompanied by the influx of Zn^{2+}, which has been suggested to be a major effector in an experimental paradigm of stroke (Koh et al. 1996). However, there is compelling evidence from a large number of different experimental models that Ca^{2+} overload and the downstream events that it triggers are the actual reason for cell demise, be it apoptotic or necrotic. Thus, the development of strategies to control cellular Ca^{2+} homeostasis remains a useful approach to prevent neurotoxicity.

References

Andreeva N, Khodorov B, Stelmashook E, Cragoe E Jr, Victorov I (1991) Inhibition of Na^+/Ca^{2+} exchange enhances delayed neuronal death elicited by glutamate in cerebellar granule cell cultures. Brain Res 548:322–325

Ankarcrona M, Dypbukt JM, Bonfoco E, Zhivotovsky B, Orrenius S, Lipton SA, Nicotera P (1995) Glutamate-induced neuronal death: a succession of necrosis or apoptosis depending on mitochondrial function. Neuron 15:961–973

Ankarcrona M, Dypbukt JM, Orrenius S, Nicotera P (1996) Calcineurin and mitochondrial function in glutamate-induced neuronal cell death. FEBS Lett 394:321–324

Bading H, Ginty DD, Greenberg ME (1993) Regulation of gene expression in hippocampal neurons by distinct calcium signaling pathways. Science 260:181–186

Bagetta G, Corasaniti T, Berliocchi L, Navarra M, Finazzi-Agrò A, Nisticö G (1995) HIV-1 gp120 produces DNA fragmentation in the cerebral cortex of rat. Biochem Biophys Res Comm 211:130–136

Barger SW, Hörster D, Furukawa K, Goodman Y, Krieglstein J, Mattson MP (1995) Tumor necrosis factor alpha and beta protect neurons aganist amyloid beta-peptide toxicity: evidence for involvement of a kappa B-binding factor and attenuation of peroxide and Ca^{2+} accumulation. Proc Natl Acad Soc USA 92:9328–32

Beilharz EJ, Williams CE, Dragunow M, Sirimanne ES, Gluckman PD (1995) Mechanisms of delayed cell death following hypoxic-ischemic injury in the immature rat: evidence for apoptosis during selective neuronal loss. Mol Brain Res 29:1–14

Beneviste H, Drejer J, Schousboe A, Diemer NH (1984) Elevation of the extracellular concentration of glutamate and aspartate in rat hippocampus during transient cerebral ischemia monitored by intracerebral microdialysis. J Neurochem 43:1369–1374

Benfenati F, Valtorta F (1995) Neuroexocytosis. Curr Top Microbiol Immunol 195:195–219

Berdichevsky E, Riveros N, Sánchez-Armáss S, Orrego F (1983) Kainate, N-methylaspartate and other excitatory amino acids increase calcium influx into rat brain cortex cells in vitro. Neurosci Lett 36:75–80

Bernardi P (1996) The permeability transition pore. Control points of a cyclosporin. A-sensitive mitochondrial channel involved in cell death. Biochim Biophys Acta 1275:5–9

Bindokas VP, Miller RJ (1995) Excitotoxic degeneration is inhibited at non-random sites in cultured rat cerebellar neurons. J Neurosci 15:6999–7011

Bonfoco E, Ceccatelli S, Manzo L, Nicotera P (1995a) Colchicine induces apoptosis in cerebellar granule cells. Exp Cell Res 218:189–200

Bonfoco E, Krainc D, Ankarcrona M, Nicotera P, Lipton SA (1995b) Apoptosis and necrosis: two distinct events induced respectively by mild and intense insults with NMDA or nitric oxide/superoxide in cortical cell cultures. Proc Natl Acad Sci USA 92:72162–72166

Bonfoco E, Leist M, Zhivotovsky B, Orrenius S, Lipton SA, Nicotera P (1996) Cytoskeletal breakdown and apoptosis elicited by NO-donors in cerebellar granule cells require NMDA-receptor activation. J Neurochem 67:2484–2493

Borst JG, Sakmann B (1996) Calcium influx and transmitter release in a fast CNS synapse. Nature 383:431–434

Bortner CD, Oldenburg NBE, Cidlowski JA (1995) The role of DNA fragmentation in apoptosis. Trends Cell Biol 5:21–26

Bouchelouche P, Belhage B, Frandsen A, Drejer J, Schousboe A (1989) Glutamate receptor activation in cultured cerebellar granule cells increases cytosolic free Ca2+ by mobilization of cellular Ca^{2+} and activation of Ca^{2+} influx. Exp Brain Res 76:281–291

Brenneman DE, Westbrook GL, Fitzgerald SP, Ennist DL, Elkins KL, Ruff MR, Pert CB (1988) Neuronal cell killing by the envelope protein of HIV and its prevention by vasoactive intestinal peptide. Nature 335:639–642

Brorson JR, Manzolillo PA, Miller RJ (1994) Ca-2+ entry via AMPA/kainate receptors and excitotoxicity in cultured cerebellar Purkinje cells. J Neurosci 14:187–197

Brorson JR, Manzolillo PA, Gibbons SJ, Miller RJ (1995) AMPA-receptor desensitization predicts the selective vulnerability of cerebellar Purkinje cells to excitotoxicity. J Neurosci 15:4515–4524

Brown DR, Schmidt B, Kretzschmar HA (1996) Role of microglia and host prion protein in neurotoxicity of a prion protein fragment. Nature 380:345–347

Buchan A, Pulsinelli WA (1990) Hypothermia but not the N-methyl-D-aspartate antagonist, MK-801, attenuates neuronal damage in gerbils subjected to transient global ischemia. J Neurosci 10:311–316

Buchan A, Li H, Pulsinelli WA (1991) The N-methyl-D-aspartate antagonist, MK-801, fails to protect against neuronal damage caused by transient, severe forebrain ischemia in adult rats. J Neurosci 11:1049–1056

Budd SL, Nicholls DG (1996a) A reevaluation of the role of mitochondria in neuronal Ca2+ homeostasis. J Neurochem 66:403–411

Budd SL, Nicholls DG (1996b) Mitochondria, calcium regulation, and acute glutamate excitotoxicity in cultured cerebellar granule cells. J Neurochem 67:2282–2291

Bührle CP, Sonnhof U (1983) The ionic mechanism of the excitatory action of glutamate upon the membranes of motoneurones of the frog. Pflügers Arch 396:154–162

Bullock R (1995) Strategies for neuroprotection with glutamate antagonists. Extrapolating from evidence taken from the first stroke and head injury studies. Ann N Y Acad Sci 765:272–278

Bullock R, Zauner A, Myseros JS, Marmarou A, Woodward JJ, Young HF (1995) Evidence for prolonged release of excitatory amino acid in serve human head trauma. Relationship to clinical events. Ann N Y Acad Sci 765:290–297

Carafoli E (1991) The Ca^{2+} pump of the plasma membrane. Physiol Rev 71:129–153

Carini R, Bellomo G, Dianzini MU, Albano E (1994) Evidence for a sodium-dependent calcium influx in isolated rat hepatocytes undergoing ATP depletion. Biochem Biophys Res Comm 202:360–366

Chard PS, Bleakman D, Savidge JR, Miller RJ (1995) Capsicin-induced neurotoxicity in cultured dorsal root ganglion neurons: involvement of calcium-activated proteases. Neurosci 65:1099–1108

Charriaut-Marlangue C, Margaill I, Borrega F, Plotkine M, Ben-Ari Y (1996) NG-Nitro-L-arginine methyl ester reduces necrotic but not apoptotic cell death induced by reversible focal ischemia in rat. Eur J Pharmacol 310:137–140

Chen H-SV, Pellegrini JW, Aggarwal SK, Lei SZ, Warach S, Jensen FE, Lipton SA (1992) Open-channel block of N-methyl-D-aspartate (NMDA) responses by memantine: therapeutic advantage against NMDA receptor-mediated neurotoxicity. J Neurosci 12:4427–4436

Cheng B, Mattson MP (1991) NGF and bFGF protect rat hippocampal and human cortical neurons against hypoglycemic damage by stabilizing calcium homeostatsis. Neuron 7:1031–1041

Cheng B, Mattson MP (1992) IGF-I and IGF-II protect cultured hippocampal and septal neurons against calcium-mediated hypoglycemic damage. J Neurosci 12:1558–1566

Cheng B, Christakos S, Mattson MP (1994) Tumor necrosis factors protect neurons against metabolic-excitotoxic insults and promote maintenance of calcium homeostasis. Neuron 12:139–153

Chittajallu R, Vignes M, Dev KK, Barnes JM, Collingridge GL, Henley JM (1996) Regulation of glutamate release by presynaptic kainate receptors in the hippocampus. Nature 379:78–81

Choi DW (1985) Glutamate neurotoxicity in cortical cell culture is calcium dependent. Neurosci Lett 58:293–297

Choi DW (1987) Ionic dependence of glutamate neurotoxicity. J Neurosci 7:369–379

Choi DW (1988a) Calcium-mediated neurotoxicity: relationship to specific channel types and role in ischemic damage. Trends Neurosci 11:465–469

Choi DW (1988b) Glutamate neurotoxicity and diseases of the nervous system. Neuron 1:623–634

Choi DW (1992) Bench to bedside: the glutamate connection. Science 258:241–243

Choi DW (1995) Calcium: still center-stage in hypoxic-ischemic neuronal death. Trends Neurosci 18:58–60

Choi DW, Rothman SM (1990) The role of glutamate neurotoxicity in hypoxic-ischemic neuronal death. Annu Rev Neurosci 13:171–182

Chou CC, Lam CY, Yung BYM (1995) Intracellular ATP is required for actinomycin D-induced apoptotic cell death in HeLa cells. Cancer Lett 96:181–187

Clawson GA, Norbeck LL, Hatem CL, Rhodes C, Amiri P, McKerrow JH, Patierno SR, Fiskum G (1992) Ca^{2+}-regulated serine protease associated with the nuclear scaffold. Cell Growth Differ 3:827–838

Collinge J, Whittington MA, Sidle KCL, Smith CJ, Palmer MS, Clarke AR, Jefferys JGR (1994) Prion protein is necessary for normal synaptic function. Nature 370:295–297

Connor JA, Wadman WJ, Hockberger PE, Wong RKS (1988) Sustained dentritic gradients of Ca^{2+} induced by excitatory amino acids in CA1 hippocampal neurons. Science 240:649–653

Cooper AJ, Wooller S, Mitchell IJ (1995) Elevated striatal Fos immunoreactivity following 6-hydrodopamine lesioning of the rat is mediated by excitatory amino acid transmission. Neurosci Lett 194:73–76

Courtney MJ, Lambert JJ, Nicholls DG (1990) The interactions between plasma membrane depolarization and glutamate receptor activation in the regulation of cytoplasmic free calcium in cultured cerebellar granule cells. J Neurosci 10:3873–3879

Cox JA, Felder CC, Henneberry RC (1990) Differential expression of excitatory amino acid receptor subtypes in cultured cerebellar neurons. Neuron 4:941–947

Coyle JT, Puttfarcken P (1993) Oxidative stress, glutamate, and neurodegenerative disorders. Science 262:689–695

Dawson TM, Steiner JP, Dawson VL, Dinerman JL, Uhl GR, Snyder SH (1993) Immunosuppressant FK 506 enhances phosphorylation of nitric oxide synthase and protects against glutamate neurotoxicity. Proc Natl Acad Sci USA 90:9808–9812

Dawson VL, Dawson TM, London ED, Bredt DS, Snyder SH (1991) Nitric oxide mediates glutamate neurotoxicity in primary cortical cultures. Proc Natl Acad Sci USA 88:6368–6371

Dawson VL, Dawson TM, Bartley DA, Uhl GR, Snyder SH (1993) Mechanisms of nitric oxide-mediated neurotoxicity in primary brain cultures. J Neurosci 13:2651–2661

De Erausquin GA, Manev H, Guidotti A, Costa E, Brooker G (1990) Gangliosides normalize distorted single-cell intracellular free Ca^{2+} dynamics after toxic doses of glutamate in cerebellar granule cells. Proc Natl Acad Sci USA 87:8017–8021

Denk W, Yuste R, Svoboda K, Tank DW (1996) Imaging calcium dynamics in dendritic spines. Curr Opin Neurobiol 6:372–378

Dienel GA (1984) Regional accumulation of calcium in postischemic rat brain. J Neurochem 43:913–925

Dowd DR, MacDonald PN, Komm BS, Haussler MR, Miesfeld RL (1992) Stable expression of the calbindin-D28 K complementary DNA interferes with the apoptotic pathway in lymphocytes. Mol Endocrinol 6:1843–1848

Drejer J, Beneviste H, Diemer NH, Schousboe A (1985) Cellular origin of ischemia-induced glutamate release from brain tissue in vivo and in vitro. J Neurochem 45:145–151

Dreyer EB, Kaiser, PK, Offermann JT, Lipton SA (1990) HIV-1 coat protein neurotoxicity prevented by calcium channel antagonists. Science 248:364–367

Dubinsky JM (1992) Examination of the role of calcium in neuronal death. Ann NY Acad Sci 679:34–42

Dubinsky JM (1993) Intracellular calcium levels during the period of delayed excitotoxicity. J Neurosci 13:623–631

Dubinsky JM, Rothman SM (1991) Intracellular calcium concentrations during <169>chemical hypoxia" and excitotoxic neuronal injury. J Neurosci 11:2545–2551

Dugan LL, Sensi SL, Canzoniero LMT, Handran SD, Rothman SM, Lin T-S, Goldberg MP, Choi DW (1995) Mitochondrial production of reactive oxygen species in cortical neurons following exposure to N-methyl-D-aspartate. J Neurosci 15:6377–6388

Dumuis A, Sebben M, Haynes L, Pin J-P, Bockaert J (1988) NMDA receptors activate the arachidonic acid cascade system in striatal neurons. Nature 336:68–70

Dumuis A, Sebben M, Fagni L, Prézeau L, Manzoni O, Cragoe EJ Jr, Bockaert J (1993) Stimulation by glutamate receptors of arachidonic acid release depends on the Na^+/Ca^{2+} exchanger in neuronal cells. Mol Pharmacol 43:976–981

Dykens JA (1994) Isolated cerebral and cerebellar mitochondria produce free radicals when exposed to elevated Ca^{2+} and Na^+: implications for neurodegeneration. J Neurochem 63:584–591

Dykens JA, Stern A, Trenkner E (1987) Mechanism of kainate toxicity to cerebellar neurons in vitro is analogous to reperfusion tissue injury. J Neurochem 49:1222–1228

Eguchi Y, Shimizu S, Tsujimoto Y (1997) Intracellular ATP levels determine cell death fate by apoptosis or necrosis. Cancer Res 57:1835–1840

Ehlers MD, Zhang S, Bernhardt JP, Huganir RL (1996) Inactivation of NMDA receptors by direct interaction of calmodulin with the NR1 subunit. Cell 84:745–755

Eimerl S, Schramm M (1994) The quantity of calcium that appears to induce neuronal death. J Neurochem 62:1223–1226

Fesus L, Thomazy V, Falus A (1987) Induction and activation of tissue transglutaminase during programmed cell death. FEBS Lett 224:104–108

Fleckenstein A (1984) Calcium antagonism: history and prospects for a multifaceted pharmacodynamic principle. In: Opie LH (ed) Calcium antagonists and cardiovascular disease. Raven, New York, pp 9–28

Forloni G, Chiesa R, Smiroldo S, Verga L, Salmona M, Tagliavini F, Angeretti N (1993) Apoptosis mediated neurotoxicity induced by chronic application of beta-amyloid fragment 25–35. Neuroreport 4:523–526

Frandsen A, Schousboe A (1991) Dantrolene prevents glutamate cytotoxicity and Ca^{2+} release from intracellular stores in cultured cerebral cortical neurons. J Neurochem 56:1075–1078

Frandsen A, Drejer J, Schousboe A (1989) Direct evidence that excitotoxicity in cultured neurons is mediated via N-methyl-D-aspartate (NMDA) as well as non-NMDA receptors. J Neurochem 53:297–299

Franklin JL, Sanz-Rodriguez C, Juhasz A, Deckwerth TL, Johnson EM Jr (1995) Chronic depolarization prevents programmed death of sympathetic neurons in vitro but does not support growth: requirement for Ca^{2+} influx but not Trk activation. J Neurosci 15:643–664

Furukawa K, Mattson MP (1995) Cytochalasins protect hippocampal neurons against amyloid beta-peptide toxicity: evidence that actin depolymerization suppresses Ca^{2+} influx. J Neurochem 65:1061–1068

Furukawa K, Barger SW, Blalock EM, Mattson MP (1996) Activation of K^+ channels and suppression of neuronal activity by secreted beta-amyloid-precursor protein. Nature 379:74–77

Galli C, Meucci O, Scorziello A, Werge TM, Calissano P, Schettini G (1995) Apoptosis in cerebellar granule cells is blocked by high KCl, forskolin, and IGF-1 through distinct mechanisms of action: the involvement of intracellular calcium and RNA synthesis. J Neurosci 15:1172–1179

Gallo V, Ciotti MT, Coletti A, Aloisi F, Levi G (1982) Selective release of glutamate from cerebellar granule cells differentiating in culture. Proc Natl Acad Sci USA 79:7919–7923

Garthwaite G, Garthwaite J (1986a) Amino acid neurotoxicity: intracellular sites of calcium accumulation associated with the onset of irreversible damage to rat cerebellar neurones in vitro. Neurosci Lett 71:53–58

Garthwaite G, Garthwaite J (1986b) Neurotoxicity of excitatory amino acid receptor agonists in rat cerebellar slices: dependence on calcium concentration. Neurosci Lett 66:193–198

Geiger JR, Melcher T, Koh DS, Sakmann B, Seeburg PH, Jonas P, Monyer H (1995) Relative abundance of subunit mRNAs determines gating and Ca^{2+} permeability of

AMPA receptors in principal neurons and interneurons in rat CNS. Neuron 15:193–204

Gelbard HA, James HJ, Sharer LR, Perry SW, Saito Y, Kazee AM, Blumberg BM, Epstein LG (1995) Apoptotic neurons in brains from paediatric patients with HIV-1 encephalitis and progressive encephalopathy. Neuropathol Appl Neurobiol 21:208–217

Ghosh A, Greenberg ME (1995) Calcium signalling in neurons: molecular mechanisms and cellular consequences. Science 268:239–247

Ghosh A, Carnahan J, Greenberg ME (1994) Requirement for BDNF in activity-dependent survival of cortical neurons. Science 263:1618–1623

Gibbons SJ, Brorson JR, Bleakman D, Chard PS, Miller RJ (1993) Calcium influx and neurodegeneration. Ann NY Acad Sci 679:22–33

Giese A, Groschup MH, Hess B, Kretzschmar HA (1995) Neuronal cell death in scrapie-infected mice is due to apoptosis. Brain Pathol 5:213–21

Gill R, Andiné P, Hillered L, Persson L, Hagberg H (1992) The effect of MK-801 on cortical spreading depression in the penumbral zone following focal ischaemia in the rat. J Cereb Blood Flow Metab 12:371–379

Ginty DD, Kornhauser JM, Thompson MA, Bading H, Mayo KE, Takahashi JS, Greenberg ME (1993) Regulation of CREB phosphorylation in the suprachiasmatic nucleus by light and circadian clock. Science 260:238–241

Giulian D, Vaca K, Noonan CA (1990) Secretion of neurotoxins by mononuclear phagocytes infected with HIV-1. Science 250:1593–1596

Goodman Y, Mattson MP (1994) Secreted forms of beta-amyloid precursor protein protect hippocampal neurons against amyloid beta-peptide-induced oxidative injury. Exp Neurol 128:1–12

Gorman AM, Scott MP, Rumsby PC, Meredith C, Griffiths R (1995) Excitatory amino acid-induced cytotoxicity in primary cultures of mouse cerebellar granule cells correlates with elevated, sustained c-fos proto-oncogene expression. Neurosci Lett 191:116–120

Grynkiewicz G, Poenie M, Tsien RY (1985) A new generation of Ca^{2+} indicators with greatly improved fluorescence properties. J Biol Chem 260:3440–3450

Gschwind M, Huber G (1995) Apoptotic cell death induced by beta-amyloid 1–42 peptide is cell type dependent. J Neurochem 65:292–300

Gu JG, Albuquerque C, Lee CJ, MacDermott AB (1996) Synaptic strengthening through activation of Ca^{2+}-permeable AMPA-receptors. Nature 381:793–795

Gunter TE, Pfeiffer DR (1990) Mechanisms by which mitochondria transport calcium. Am J Physiol 258:C755–C786

Gunter TE, Gunter KK, Sheu S, Gavin CE (1994) Mitochondrial calcium transport: physiological and pathological relevance. Am J Physiol 267:C313–C339

Hahn JS, Aizenman E, Lipton SA (1988) Central mammalian neurons normally resistant to glutamate toxicity are made sensitive by elevated extracellular Ca^{2+}: toxicity is blocked by the N-methyl-D-aspartate antagonist MK-801. Proc Natl Acad Sci USA 85:6556–6560

Hajos F, Garthwaite G, Garthwaite J (1986a) Reversible and irreversible neuronal damage caused by excitatory amino acid analogues in rat cerebellar slices. Neurosci 18:417–436

Hajos F, Garthwaite G, Garthwaite J (1986b) Ionic requirements for neurotoxic effects of excitatory amino acid analogues in rat cerebellar slices. Neurosci 18:437–447

Hardingham GE, Chawla S, Johnson CM, Bading H (1997) Distinct functions of nuclear and cytoplasmic calcium in the control of gene expression. Nature 385:260–265

Hartley A, Stone JM, Heron C, Cooper JM, Schapira AHV (1994) Complex I inhibitors induce dose-dependent apoptosis in PC12 cells: relevance to Parkinson's disease. J Neurochem 63:1987–1990

Hartley DM, Choi DW (1989) Delayed rescue of N-methyl-D-aspartate receptor-mediated neuronal injury in cortical culture. J Pharmacol Exp Therap 250:752–758

Heinemann U, Pumain R (1980) Extracellular calcium activity changes in cat sensorimotor cortex induced by iontophoretic application of aminoacids. Exp Brain Res 40:247–250

Hernández-Cruz A, Sala F, Adams PR (1990) Subcellular calcium transients visualized by confocal microscopy in a voltage-clamped vertebrate neuron. Science 247:858–862

Hewish DR, Burgoyne LA (1973) Chromatin sub-structure. The digestion of chromatin DNA at regularly spaced sites by a nuclear deoxyribonuclease. Biochem Biophys Res Comm 52:504–510

Hollmann M, Heinemann S (1994) Cloned glutamate receptors. Annu Rev Neurosci 17:31–108

Holzwarth JA, Gibbons SJ, Brorson JR, Philipson LH, Miller RJ (1994) Glutamate receptor agonists stimulate diverse calcium responses in different types of cultured rat cortical glial cells. J Neurosci 14:1879–1891

Huang L-YM, Neher E (1996) Ca^{2+}-dependent exocytosis in the somata of dorsal root ganglion neurons. Neuron 17:135–145

Huang Z, Huang PL, Panahian N, Dalkara T, Fishman MC, Moskowitz MA (1994) Effects of cerebral ischemia in mice deficient in neuronal nitric oxide synthase. Science 265:1883–1885

Iadecola C (1997) Bright and dark sides of nitric oxide in ischemic brain injury. Trends Neurosci 20:132–139

Ishimaru H, Katoh A, Suzuki H, Fukuta T, Kameyama T, Nabeshima T (1992) Effects of N-methyl-D-aspartate receptor antagonists on carbon monoxide-induced brain damage in mice. J Pharmacol Exp Ther 261:349–352

James P, Vorherr T, Carafoli E (1995) Calmodulin-binding domains: just two faced or multi-faceted?. Trends Biochem Sci 20:38–42

Johnson BD, Byerly L (1993) A cytoskeletal mechanism for Ca^{2+} channel metabolic dependence and inactivation by intracellular Ca^{2+}. Neuron 10:797–804

Johnson EM, Koike T, Franklin J (1992) A "calcium set-point hypothesis" of neuronal dependence on neurotrophic factor. Exp Neurol 115:163–166

Johnson EM, Deckwerth TL (1993) Molecular mechanisms of developmental neuronal death. Annu Rev Neurosci 16:31–46

Kaiser-Petito C, Roberts B (1995) Evidence of apoptotic cell death in HIV encephalitis. Am J Pathol 146:1121–1130

Kass IS, Lipton P (1986) Calcium and long-term transmission damage following anoxia in dentate gyrus and CA1 regions of the rat hippocampal slice. J Physiol 378:313–334

Kiedrowski L, Costa E (1995) Glutamate-induced destabilization of intracellular calcium concentration homeostasis in cultured cerebellar granule cells: role of mitochondria in calcium buffering. Mol Pharmacol 47:140–147

Kiedrowski L, Brooker G, Costa E, Wroblewski JT (1994) Glutamate impairs neuronal calcium extrusion while reducing sodium gradient. Neuron 12:295–300

Kluck RM, McDougall CA, Harmon BV, Halliday JW (1994) Calcium chelators induce apoptosis – evidence that raised intracellular ionised calcium is not essential for apoptosis. Biochim Biophys Acta 1223:247–254

Kluck RM, Bossy-Wetzel E, Green DR, Newmeyer DD (1997) The release of cytochrome c from mitochondria: a primary site for bcl-2 regulation of apoptosis. Science 275:1132–1136

Koh J-Y, Choi DW (1988) Vulnerability of cultured cortical neurons to damage by excitotoxins: differential susceptibility of neurons containing NADPH-diaphorase. J Neurosci 8:2153–2163

Koh J-Y, Peters S, Choi DW (1986) Neurons containing NADPH-diaphorase are selectively resistant to quinolinate toxicity. Science 234:73–76

Koh J-Y, Suh SW, Gwang BJ, He YY, Hsu CY, Choi DW (1996) The role of zinc in selective neuronal death after transient global cerebral ischemia. Science 272:1013–1016

Koike T, Martin DP, Johnson EM Jr (1989) Role of Ca^{2+} channels in the ability of membrane depolarization to prevent neuronal death induced by trophic-factor deprivation: evidence that levels of internal Ca^{2+} determine nerve growth factor dependence of sympathetic ganglion cells. Proc Natl Acad Sci USA 86:6421–6425

Kristián T, Katsura K-I, Gidö G, Siesjö BK (1994) The influence of pH on cellular calcium influx during ischemia. Brain Res 641:295–302

Lafon-Cazal M, Clucasi M, Gaven F, Pietri S, Bockaert J (1993) Nitric oxide, superoxide and peroxynitrite: putative mediators of NMDA-induced cell death in cerebellar granule cells. Neuropharmacology 32:1259–1266

Lassmann H, Bancher C, Breitschopf H, Wegiel J, Bobinski M, Jellinger K, Wisniewski HM (1995) Cell death in Alzheimer's disease evaluated by DNA fragmentation in situ. Acta Neuropathol 89:35–41

Le W-D, Colom LV, Xie W-J, Smith RG, Alexianu M, Appel SH (1995) Cell death induced by beta-amyloid 1–40 in MES 23.5 hybrid clone: the role of nitric oxide and NMDA-gated channel activation leading to apoptosis. Brain Res 686:49–60

Lechleiter J, Girard S, Peralta E, Clapham D (1991) Spiral calcium wave propagation and annihilation in xenus laevis oocytes. Science 252:123–126

Lee KS, Frank S, Vanderklish P, Arai A, Lynch G (1991) Inhibition of proteolysis protects hippocampal neurons from ischemia. Proc Natl Acad Sci USA 88:7233–7237

Leist M, Fava E, Montecucco C, Nicotera P (1997a) Peroxynitrite and NO-donors induce neuronal apoptosis by eliciting autocrine excitotoxicity. Eur J Neurosci 9:1488–1498

Leist M., Single B, Castoldi AF, Kühnle S, Nicotera P (1997b) Intracellular ATP concentration: a switch deciding between apoptosis and necrosis. J Exp Med 185:1481–1486

Leonard JP, Salpeter MM (1979) Agonist-induced myopathy at the neuromuscular junction is mediated by calcium. J Cell Biol 82:811–819

Levi G, Aloisi F, Ciotti MT, Gallo V (1984) Autoradiographic localization and depolarization-induced release of acidic amino acids in differentiating cerebellar granule cell cultures. Brain Res 290:77–86

Levi G, Patrizio M, Gallo V (1991) Release of endogenous and newly synthesized glutamate and of other amino acids induced by non-N-methyl-D-aspartate receptor activation in cerebellar granule cell cultures. J Neurochem 56:199–206

Li Y, Chopp M, Jiang N, Zaloga C (1995a) In situ detection of DNA fragmentation after focal cerebral ischemia in mice. Mol Brain Res 28:164–168

Li Y, Chopp M, Jiang N, Yao F, Zaloga C (1995b) Temporal profile of in situ DNA fragmentation after transient middle cerebral artery occlusion in the rat. J Cereb Blood Flow Metab 15:389–397

Li Y, Chopp M, Jiang N, Zhang ZG, Zaloga C (1995c) Induction of DNA fragmentation after 10 to 120 minutes of focal cerebral ischemia in rats. Stroke 26:1252–1258

Li Y, Sharov VG, Jiang N, Zaloga C, Sabbah HN, Chopp M (1995d) Ultrastructural and light microscopic evidence of apoptosis after middle cerebral artery occlusion in the rat. Am J Pathol 146:1045–1051

Lieberman DN, Mody I (1994) Regulation of NMDA channel function by endogenous Ca^{2+}-dependent phosphatase. Nature 369:235–239

Linnik MD, Miller JA, Sprinkle-Cavallo J, Mason PJ, Thompson FY, Montgomery LR, Schroeder KK (1995) Apoptotic DNA fragmentation in the rat cerebral cortex induced by permanent middle cerebral artery occlusion. Mol Brain Res 32:116–124

Lipton SA (1992a) Models of neuronal injury in AIDS: another role for the NMDA receptor?. Trends Neurosci 15:75–79

Lipton SA (1992b) 7-Chlorokynurenate ameliorates neuronal injury mediated by HIV envelope protein gp120 in rodent retinal cultures. Eur J Neurosci 4:1411–1415

Lipton SA (1992c) Requirement for macrophages in neuronal injury induced by HIV envelope protein gp120. Neuroreport 3:913–915

Lipton SA, Gendelman HE (1995) Dementia associated with the acquired immunodeficiency syndrome. New Engl J Med 332:934–940

Lipton SA, Rosenberg PA (1994) Excitatory amino acids as a final common pathway for neurologic disorders. New Engl J Med 330:613–622

Lipton SA, Sucher NJ, Kaiser PK, Dreyer EB (1991) Synergistic effects of HIV coat protein and NMDA receptor-mediated neurotoxicity. Neuron 7:111–118

Liu X, Kim CN, Yang J, Jemmerson R, Wang X (1996) Induction of apoptotic program in cell-free extracts: requirement for dATP and cytochrome c. Cell 86:147–157

Lu YM, Yin HZ, Chiang J, Weiss JH (1996) Ca^{2+}-permeable AMPA/kainate and NMDA channels: high rate of Ca^{2+} influx underlies potent induction of injury. J Neurosci 16:5457–5465

Lucas DR, Newhouse JP (1957) The toxic effect of sodium L-glutamate on the inner layers of the retina. Arch Ophthalmol 58:193–201

Maiese K, Swiriduk M, TenBroeke M (1996) Cellular mechanisms of protection by metabotropic glutamate receptors during anoxia and nitric oxide toxicity. J Neurochem 66:2419–2428

Malgaroli A, Tsien RW (1992) Glutamate-induced long-term potentiation of the frequency of miniature synaptic currents in cultured hippocampal neurons. Nature 357:134–139

Malgaroli A, Milani D, Meldolesi J, Pozzan T (1987) Fura-2 measurement of cytosolic free Ca^{2+} in monolayers and suspensions of various types of animal cells. J Cell Biol 105:2145–2155

Manev H, Favaron M, Guidotti A, Costa E (1989) Delayed increase of Ca^{2+} influx elicited by glutamate: role in neuronal death. Mol Pharmacol 36:106–112

Marcaida G, Minana M-D, Grisolía S, Felipo V (1995) Lack of correlation between glutamate-induced depletion of ATP and neuronal death in primary cultures of cerebellum. Brain Res 695:146–150

Marcaida G, Kosenko E, Minana M-D, Grisolía S, Felipo V (1996) Glutamate induces a calcineurin-mediated dephosphorylation of Na^+, K^+-ATPase that results in its activation in cerebellar neurons in culture. J Neurochem 66:99–104

Marciani MG, Louvel J, Heinemann U (1982) Aspartate-induced changes in extracellular free calcium in in vitro hippocampal slices of rats. Brain Res 238:272–277

Marin P, Quignard J-F, Lafon-Cazal M, Bockaert J (1993) Non-classical glutamate receptors, blocked by both NMDA and non-NMDA antagonists, stimulate nitric oxide production in neurons. Neuropharmacol 32:29–36

Mattson MP (1990) Antigenic changes similar to those seen in neurofibrillary tangles are elicited by glutamate and Ca^{2+} influx in cultured hippocampal neurons. Neuron 2:105–117

Mattson MP, Guthrie PB, Kater SB (1989a) A role for Na+-dependent Ca^{2+} extrusion in protection against neuronal excitotoxicity. FASEB J 3:2519–2526

Mattson MP, Guthrie PB, Hayes BC, Kater SB (1989b) Roles for mitotic history in the generation and degeneration of hippocampal neuroarchitecture. J Neurosci 9:1223–1232

Mattson MP, Rychlik B, Chu C, Christakos S (1991) Evidence for calcium-reducing and excito-protective roles for the calcium-binding protein Calbindin-D28 k in cultured hippocampal neurons. Neuron 6:41–51

Mattson MP, Cheng B, Davis D, Bryant K, Lieberburg I, Rydel RE (1992) Beta-amyloid peptides destabilize calcium homeostasis and render human cortical neurons vulnerable to excitotoxicity. J Neurosci 12:376–389

Mattson MP, Kumar KN, Wang H, Cheng B, Michaelis EK (1993a) Basic FGF regulates the expression of a functional 71 kDa NMDA receptor protein that mediates calcium influx and neurotoxicity in hippocampal neurons. J Neurosci 13:4575–4588

Mattson MP, Tomaselli KJ, Rydel RE (1993b) Calcium-destabilizing and neurodegenerative effects of aggregated beta-amyloid peptide are attenuated by basic FGF. Brain Res 621:35–49

Mattson MP, Zhang Y, Bose S (1993c) Growth factors prevent mitochondrial dysfunction, loss of calcium homeostasis, and cell injury, but not ATP depletion in hippocampal neurons deprived of glucose. Exp Neurol 121:1–13

Mattson MP, Barger SW, Cheng B, Lieberburg I, Smith-Swintowsky VL, Rydel RE (1993d) Beta-amyloid protein metabolites and loss of neuronal Ca^{2+} homeostasis in Alzheimer's disease. Trends Neurosci. 16:409–414

Mattson MP, Rydel RE, Lieberburg I, Smith-Swintosky VL (1993e) Altered calcium signaling and neuronal injury: stroke and Alzheimer's disease as examples. Ann N Y Acad Sci 679:1–21

Mattson MP, Chen B, Smith-Swintosky VL (1993f) Neurotrophic factor mediated protection from excitotoxicity and disturbances in calcium and free radical metabolism. Sem Neurosci 5:295–307

Mattson MP, Cheng B, Culwell AR, Esch FS, Lieberburg I, Rydel RE (1993g) Evidence for excitoprotective and intraneuronal calcium-regulating roles for secreted forms of the beta-amyloid precursor protein. Neuron 10:243–254

Mattson MP, Barger SW, Begley JG, Mark RJ (1995) Calcium, free radicals, and excitotoxic neuronal death in primary cell culture. Meth Cell Biol 46:187–216

Meffert MK, Premack BA, Schulman H (1994) Nitric oxide stimulates calcium-independent synaptic vesicle release. Neuron 12:1235–1244

Meffert MK, Calakos NC, Scheller RH, Schulman H (1996) Nitric oxide modulates synaptic vesicle docking/fusion reactions. Neuron 16:1229–1236

Meldrum B, Garthwaite J (1990) Excitatory amino acid neurotoxicity and neurodegenerative disease. Trends Pharmacol Sci 11:379–387

Melino G, Annicchiarico-Petruzzelli M, Piredda L, Candi E, Gentile V, Davies PJA, Piacentini M (1994) Tissue transglutaminase and apoptosis: sense and antisense transfection studies with human neuroblastoma cells. Mol Cell Biol 14:6584–6596

Michaels RL, Rothman SM (1990) Glutamate neurotoxicity in vitro: antagonist pharmacology and intracellular calcium concentrations. J Neurosci 10:283–292

Milani D, Guidolin D, Facci L, Pozzan T, Buso M, Leon A, Skaper SD (1991) Excitatory amino acid-induced alterations of cytoplasmic free Ca^{2+} in individual cerebellar granule neurons: role in neurotoxicity. J Neurosci Res 28:434–441

Miljanich GP, Ramachandran J (1995) Antagonists of neuronal calcium channels: structure, function, and therapeutic implications. Annu Rev Pharmacol Toxicol 35:707–734

Mills JC, Nelson D, Erecinska M, Pittman RN (1995) Metabolic and energetic changes during apoptosis in neural cells. J Neurochem 65:1721–1730

Mitchell IJ, Lawson S, Moser B, Laidlaw SM, Cooper AJ, Walkinshaw G, Waters CM (1994) Glutamate-induced apoptosis results in a loss of striatal neurons in the parkinsonian rat. Neurosci 63:1–5

Mogensen HS, Hack N, Balázs R, Jorgensen OS (1994) The survival of cultured mouse cerebellar granule cells is not dependent on elevated potassium-ion concentration. Int J Devl Neurosci 12:451–460

Monyer H, Sprengel R, Schoepfer R, Herb A, Higuchi M, Lomeli H, Burnashev N, Sakmann B, Seeburg PH (1992a) Heteromeric NMDA receptors: molecular and functional distinction of subtypes. Science 256:1217–1221

Monyer H, Giffard RG, Hartley DM, Dugan LL, Goldberg, MP, Choi DW (1992b) Oxygen or glucose deprivation-induced neuronal injury in cortical cell cultures is reduced by tetanus toxin. Neuron 8:967–973

Morgan JI, Curran T (1986) Role of ion flux in the control of c-fos expression. Nature 322:552–555

Mukhin A, Fan L, Faden AI (1996) Activation of metabotropic glutamate receptors subtype mGluR1 contributes to post-traumatic neuronal injury. J Neurosci 16:6012–6020

Müller WEG, Schröder HC, Ushijima H, Dapper J, Bormann J (1992) gp120 of HIV-1 induces apoptosis in rat cortical cell cultures: prevention by memantine. Eur J Pharmacol 226:209–214

Müller WEG, Ushijima H, Schröder HC, Forrest JMS, Schatton WFH, Rytik PG, Heffner-Lauc M (1993) Cytoprotective effect of NMDA receptor antagonists on prion protein (PrionSc)-induced toxicity in rat cortical cell cultures. Eur J Pharmacol 246:261–267

Murphy SN, Miller RJ (1989a) Regulation of Ca^{2+} influx intro strital neurons by kainic acid. J Pharmacol Exp Ther 249:184–193

Murphy SN, Miller RJ (1989b) Two distinct quisqualate receptors regulate Ca^{2+} homeostasts in hippocampal neurons in vitro. Mol Pharmacol 35:671–680

Murphy SN, Thayer SA, Miller RJ (1987) The effects of excitatory amino acid on intracellular calcium in single mouse striatal neurons in vitro. J Neurosci 7:4145–4158

Myseros JS, Bullock R (1995) The rationale for glutamate antagonists in the treatment of traumatic brain injury. Ann N Y Acad Sci 765:262–271

Nedergaard M (1994) Direct signaling from astrocytes to neurons in cultures of mammalian brain cells. Science 263:1768–1771

Newmeyer DD, Farschon DM, Reed JC (1994) Cell-free apoptosis in xenopus egg extracts: inhibition by bcl-2 and requirement for an organelle fraction enriched in mitochondria. Cell 79:353–364

Nicholls D, Attwell D (1990) The release and uptake of excitatory amino acids. Trends Pharmacol Sci 11:462–468

Nicotera P, Rossi A (1993) Molecular mechanisms of metal neurotoxicity. J Trace Elem Electrolytes Health Dis 7:254–256

Nicotera P, McConkey DJ, Jones DP, Orrenius S (1989) ATP stimulates Ca^{2+} uptake and increases the free Ca^{2+} concentration in isolated rat liver nuclei. Proc Natl Acad Sci USA 86:453–457

Nicotera P, Orrenius S, Nilsson T, Berggren P-O (1990) An inositol 1,4,5-triphosphate-sensitive Ca^{2+} pool in liver nuclei. Proc Natl Acad Sci USA 87:6858–6862

Nicotera P, Bellomo G, Orrenius S (1992) Calcium-mediated mechanisms in chemically induced cell death. Annu Rev Pharmacol Toxicol 32:449–470

Nicotera P, Zhivotovsky B, Orrenius S (1994) Nuclear calcium transport and the role of calcium in apoptosis. Cell Calcium 16:279–288

Novelli A, Reilly JA, Lysko PG, Henneberry RC (1988) Glutamate becomes neurotoxic via the N-methyl-D-aspartate receptor when intracellular energy levels are reduced. Brain Res 451:205–212

Nuccitelli R (ed)(1994) A practical guide to the study of calcium in living cells. (Methods in cell biology, vol 40) Academic, San Diego

Olney JW (1969) Glutamate-induced retinal degeneration in neonatal mice. Electron microscopy of the acutely evolving lesion. J Neuropathol Exp Neurol 28:455–474

O'Malley DM (1994) Calcium permeability of the neuronal nuclear envelope: evaluation using confocal volumes and intracellular perfusion. J Neurosci 14:5741–5758

Orrenius S, McConkey DJ, Bellomo G, Nicotera P (1989) Role of Ca^{2+} in toxic cell killing. Trends Pharmacol Sci 10:281–285

Orrenius S, Burkitt MJ, Kass GEN, Dypbukt JM, Nicotera P (1992) Calcium ions and oxidative injury. Ann Neurol 32:S33–S42

Palaiologos G, Hertz L, Schousboe A (1989) Role of aspartate aminotransferase and mitochondrial dicarboxylate transport for release of endogenously and exogenously supplied neurotransmitter in glutamatergic neurons . Neurochem Res 14:35996366

Pastuszko A, Wilson DF (1985) Kainate-induced uptake of calcium by synaptosomes from rat brain. FEBS Lett 192:61–65

Piacentini M, Annicchiarico-Petruzzelli M, Oliverio S, Piredda L, Biedler JL, Melino G (1992) Phenotype-specific "tissue" transglutaminase regulation in human neuroblastoma cells in response to retinoic acid: correlation with cell death by apoptosis. Int J Cancer 52:271–278

Piani D, Fontana A (1994) Involvement of the cystine transport system xc-in the macrophage-induced glutamate-dependent cytotoxicity to neurons. J Immunol 152:3578–3585

Pike CJ, Burdick D, Walencewicz AJ, Glabe CG, Cotman CW (1993) Neurodegeneration induced by beta-amyloid peptides in vitro: the role of peptide assembly state. J Neurosci 13:1676–1687

Pizzi M, Consolandi O, Memo M, Spano P (1996) Activation of multiple metabotropic glutamate receptor subtypes prevents NMDA-induced excitotoxicity in rat hippocampal slices. Eur J Neurosci 8:1516–1521

Pollard H, Charriaut-Marlangue C, Centagrel A, Represa A, Robain O, Moreau J, Ben-Ari Y (1994) Kainate-induced apoptotic cell death in hippocampal neurons. Neurosci 63:7–18

Portera-Cailliau C, Hedreen JC, Price DL, Koliatsos VE (1995) Evidence for apoptotic cell death in Huntington disease and excitotoxic animal models. J Neurosci 15:3775–3787

Pozzan T, Rizzuto R, Volpe P, Meldolesi J (1994) Molecular and cellular physiology of intracellular calcium stores. Physiol Rev 74:595–636

Prusiner SB (1996) Human prion diseases and neurodegeneration. Curr Top Microbiol Immunol 207:1–17

Przywara DA, Bhave SV, Bhave A, Wakade TD, Wakade AR (1991) Stimulated rise in neuronal calcium is faster and greater in the nucleus than the cytosol. FASEB J 5:217–222

Randall RD, Thayer SA (1992) Glutamate-induced calcium transient triggers delayed calcium overload and neurotoxicity in rat hippocampal neurons. J Neurosci 12:1882–1895

Ratan RR, Murphy TH, Baraban JM (1994) Oxidative stress induces apoptosis in embryonic cortical neurons. J Neurochem 62:376–379

Rego AC, Santos MS, Oliveira CR (1996) Oxidative stress, hypoxia, and ischemia-like conditions increase the release of endogenous amino acids by distinct mechanisms in cultured retinal cells. J Neurochem 66:2506–2516

Reuter H (1995) Measurements of exocytosis from single presynaptic nerve terminals reveal heterogenous inhibition by Ca^{2+}-channel blockers. Neuron 14:773–779

Reuter H, Porzig H (1995) Localization and functional significance of the Na^+/Ca^{2+} exchanger in presynaptic boutons of hippocampal cells in culture. Neuron 15:1077–1084

Reynolds IJ, Hastings TG (1995) Glutamate induces the production of reactive oxygen species in cultured forebrain neurons following NMDA receptors activation. J Neurosci 15:3318–3327

Richter C, Gogvadze V, Schlapbach R, Schweizer M, Schlegel J (1994) Nitric oxide kills hepatocytes by mobilizing mitochondrial calcium. Biochem Biophys Res Commun 205:1143–1150

Rizzuto R, Brini M, Bastianutto C, Marsault R, Pozzan T (1995) Photoprotein-mediated measurement of calcium ion concentration in mitochondria of living cells. Methods Enzymol 260:417–428

Rosenmund C, Westbrook GL (1993) Calcium-induced actin depolymerization reduces NMDA channel activity. Neuron 10:805–814

Rossi AD, Larsson O, Manzo L, Orrenius S, Vather M, Berggren P-O, Nicotera P (1993) Modification of Ca^{2+} signaling by inorganic mercury in PC12 cells. FASEB J 7:1507–1514

Rossi AD, Viviani B, Zhivotovsky B, Manzo L, Orrenius S, Vahter M, Nicotera P (1997) Inorganic mercury modifies calcium-$^{2+}$ signalling, triggers apoptosis and potentiates NMDA toxicity in neural cells. Cell Death Differ 4:317–324

Rothman SM (1983) Synaptic activity mediates death of hypoxic neurons. Science 220:536–537

Rothman SM (1984) Synaptic release of excitatory amino acid neurotransmitter mediates anoxic neuronal death. J Neurosci 4:1884–1891

Rothman SM, Olney JW (1995) Excitotoxicity and the NMDA receptor – still lethal after eight years. Trends Neurosci 18:57–58

Rothman SM, Thurston JH, Hauhart RE (1987) Delayed neurotoxicity of excitatory amino acids in vitro. Neurosci 22:471–480

Rothstein JD, Dykes-Hoberg M, Pardo CA, Bristol LA, Jin L, Kuncl PW, Kanai Y, Hediger MA, Wang Y, Schielke JP, Welty DF (1996) Knockout of glutamate transporters reveals a major role for astroglial transport in excitotoxicity and clearance of glutamate. Neuron 16:576–586

Rutter GA, Theler JM, Murgia M, Wollheim CB, Pozzan T, Rizzuto R (1993) Stimulated Ca^{2+} influx raises mitochondrial free Ca^{2+} to supramicromolar levels in a pancreatic beta-cell line. Possible role in glucose and agonist-induced insulin secretion. J Biol Chem 268:22385–22390

Saido TC, Sorimachi H, Suzuki K (1994) Calpain: new perspectives in molecular diversity and physiolgical-pathological involvement. FASEB J 8:814–822

Sandberg M, Butcher SP, Hagberg H (1986) Extracellular overflow of neuroactive amino acids during severe insulin-induced hypoglycemia: in vivo dialysis of the rat hippocampus. J Neurochem 47:178–184

Sandberg M, Butcher SP, Hagberg H (1986) Extracellular overflow of neuroactive amino acids during severe insulin-induced hypoglycemia: in vivo dialysis of the rat hippocampus. J Neurochem 47:178–184

Schanne FAX, Kane AB, Young EE, Farber JL (1979) Calcium dependence of toxic cell death. Science 206:700–702

Schiavo G, Poulain B, Benfenati F, DasGupta BR, Montecucco C (1993) Novel targets and catalytic activities of bacterial protein toxins. Trends Microbiol 1:170–174

Schiavo G, Rossetto O, Tonello F, Montecucco C (1995) Intracellular targets and metalloprotease activity of tetanus and botulism neurotoxins. Curr Top Microbiol Immunol 195:257–274

Schinder AF, Olson EC, Spitzer NC, Montal M (1996) Mitochondrial dysfunction is a primary event in glutamate neurotoxicity. J Neurosci 16:6125–133

Schulz JB, Huang PL, Matthews RT, Passov D, Fishman MC, Beal MF (1996) Striatal malonate lesions are attenuated in neuronal nitric oxide synthase knockout mice. J Neurochem 67:430–433

Scorziello A, Meucci O, Florio T, Fattore M, Forloni G, Salmona M, Schettini G (1996) beta25–35 alters calcium homostasis and induces neurotoxicity in cerebellar granule cells. J Neurochem 66:1995–2003

Seeburg PH (1993) The TINS/TiPS lecture. The molecular biology of mammalian glutamate receptor channels. Trends Neurosci 16:359–365

Sharkey J, Butcher SP (1994) Immunophilins mediate the neuroprotective effects of FK506 in focal cerebral ischaemia. Nature 371:336–339

Sheardown MJ, Nielsen EO, Hansen EJ, Jacobsen P, Honoré T (1990) 2,3-dihydroxy-6-nitro-7-sulfamoyl-benzo(F)quinoxaline: a new neuroprotectant for cerebral ischemia. Science 247:571–574

Shelanski ML (1990) Intracellular ionic calcium and the cytoskeleton in living cells. Ann N Y Acad Sci 568:121–124

Sheng Z-H, Rettig J, Cook T, Catterall WA (1996) Calcium-dependent interaction of N-type calcium channels with the synaptic core complex. Nature 379:451–455

Shibasaki F, Price ER, Milan D, McKeon F (1996) Role of kinases and the phosphatase calcineurin in the nuclear shuttling of transcription factor NF-AT4. Nature 382:370–373

Shimizu S, Eguchi Y, Kamiike W, Itoh Y, Hasegawa J, Yamabe K, Otsuki Y, Matsuda H, Tsujimoto Y (1996) Induction of apoptosis as well as necrosis by hypoxia and predominant prevention of apoptosis by Bcl-2 and Bcl-xL. Cancer Res 56:2161–2166

Siesjö BK (1981) Cell damage in the brain: a speculative synthesis. J Cereb Blood Flow Metab 1:155–185

Siesjö BK, Bengtsson F (1989) Calcium fluxes, calcium antagonists, and calcium-related pathology in brain ischemia, hypoglycemia, and spreading depression: a unifying hypothesis. J Cereb Blood Flow Metab 9:127–140

Siman R, Noszek JC (1988) Excitatory amino acids activate calpain I and induce structural protein breakdown in vivo. Neuron 1:279–287

Simon RP, Swan JH, Griffiths T, Meldrum BS (1984a) Blockade of N-methyl-D-aspartate receptors may protect against ischemic damage in the brain. Science 226:850–852

Simon RP, Griffiths T, Evan MC, Swan JH, Meldrum BS (1984b) Calcium overload in selectively vulnerable neurons of the hippocampus during and after ischemia: an electron microscopic study in the rat. J Cereb Blood Flow Metab 4:350–361

Smale G, Nichols NR, Brady DR, Finch CE, Horton WE Jr (1995) Evidence for apoptotic cell death in Alzheimer's disease. Exp Neurol 133:225–230

Smeyne RJ, Vendrell M, Hayward M, Baker SJ, Miao GG, Schilling K, Robertson LM, Curran T, Morgan JI (1993) Continuous c-fos expression precedes programmmed cell death in vivo. Nature 363:166–169

Snyder SH, Sabatini DM (1995) Immunophilins and the nervous system. Nature Med 1:32–37

Stehno-Bittel L, Perez-Terzic C, Clapham DE (1995) Diffusion across the nuclear envelope inhibited by depletion of the nuclear Ca^{2+} store. Science 270:1835–1838

Sucher NJ, Lei SZ, Lipton SA (1991a) Calcium channel antagonists attenuate NMDA receptor-mediated neurotoxicity of retinal ganglion cells in culture. Brain Res 297:297–302

Sucher NJ, Aizenman E, Lipton SA (1991b) N-Methyl-D-aspartate antagonists prevent kainate neurotoxicity in rat retinal ganglion cells in vitro. J Neurosci 11:966–971

Susin SA, Zamzami N, Castedo M, Hirsch T, Marchetti P, Macho A, Daugas E, Geuskens M, Kroemer G (1996) Bcl-2 inhibits the mitochondrial release of an apoptogenic protease. J Exp Med 184:1331–1341

Svoboda K, Denk W, Kleinfeld D, Tank DW (1997) In vivo dentritic calcium dynamics in neocortical pyramidal neurons. Nature 385:161–165

Timmerman LA, Clipstone NA, Ho SN, Northrop JP, Crabtree GR (1996) Rapid shuttling of NF-AT in discrimination of Ca^{2+} signals and immunosuppression. Nature 383:837–840

Tingley WG, Roche KW, Thompson AK, Huganir RL (1993) Regulation of NMDA receptor phosphorylation by alternative splicing of the C-terminal domain. Nature 364:70–73

Toggas SM, Masliah E, Rockenstein EM, Rall GF, Abraham CR, Mucke L (1994) Central nervous system damage produced by expression of the HIV-1 coat protein gp120 in transgenic mice. Nature 367:188–193

Tong G, Shepherd D, Jahr CE (1995) Synaptic desensitization of NMDA receptors by calcineurin. Science 267:1510–1512

Traystman RJ, Kirsch JR, Koehler RC (1991) Oxygen radical mechanisms of brain injury following ischemia and reperfusion. J Appl Physiol 71:1185–1195

Trump BF, Berezesky IK (1995) Calcium-mediated cell injury and cell death. FASEB J 9:219–228

Turski L, Bressler K, Rettig K-J, Löschmann P-A, Wachtel H (1991) Protection of substantia nigra from MPP^+ neurotoxicity by N-methyl-D-aspartate antagonists. Nature 349:414–418

Tymianski M, Charlton MP, Carlen PL, Tator CH (1993a) Source specificity of early calcium neurotoxicity in cultured embryonic spinal neurons. J Neurosci 13:2085–2104

Tymianski M, Wallace MC, Spigelman I, Uno M, Carlen PL, Tator CH, Charlton MP (1993b) Cell-permanent Ca^{2+} chelators reduce early excitotoxic and ischemic neuronal injury in vitro and in vivo. Neuron 11:221–235

Tymianski M, Charlton MP, Carlen PL, Tator CH (1994) Properties of neuroprotective cell-permeant Ca^{2+} chelators: effects on $[Ca^{2+}]_i$ and glutamate neurotoxicity in vitro. J Neurophysiol 12:1973–1992

Valentino K, Newcomb R, Gadbois T, Singh T, Bowersox S, Bitner S, Justice A, Yamashiro D, Hoffman BB, Ciaranello R, Miljanich G, Ramachandran J (1993) A selective N-type calcium channel antagonist protects against neuronal loss after global cerebral ischemia. Proc Natl Acad Sci USA 90:7894–7897

Van Vlient BJ, Sebben M, Dumuis A, Gabrion J, Bockaert J, Pin J-P (1989) Endogenous amino acid release from cultured cerebellar neuronal cells: effect of tetanus toxin on glutamate release. J Neurochem 52:1229–1239

Verkhratsky A, Shmigol A (1996) Calcium-induced calcium release in neurons. Cell Calcium 19:1–14

Vilbulsreth S, Hefti F, Ginsberg MD, Dietrich WD, Busto P (1987) Astrocytes protect cultured neurons from degeneration induced by anoxia. Brain Res 422:303–311

Viviani B, Rossi AD, Chow SC, Nicotera P (1995) Organotin compounds induce calcium overload and apoptosis in PC12 cells. Neurotoxicology 16:19–26

Viviani B, Rossi AD, Chow SC, Nicotera P (1996) Triethyltin interferes with Ca^{2+} signaling and potentiates norepinephrine release in PC12 cells. Toxicol Appl Pharmacol 140:289–295

Volterra A, Trotti D, Cassutti P, Tromba C, Salvaggio A, Melcangi RC, Racagni G (1992) High sensitivity of glutamate uptake to extracellular free arachidonic acid levels in rat cortical synaptosomes and astrocytes. J Neurochem 59:600–606

Wahlestedt C, Golanov E, Yamamoto S, Yee F, Ericson H, Yoo H, Inturrisi CE, Reis DJ (1993) Antisense oligodeoxynucleotides to NMDA-R1 receptor channel protect cortical neurons from excitotoxicity and reduce focal ischemia infarctions. Nature 363:260–263

Wang KKW, Yuen P-W (1997) Development and therapeutic potential of calpain inhibitors. Adv Pharmacol 37:117–153

Wang YT, Salter MW (1994) Regulation of NMDA receptors by tyrosine kinases and phosphatases. Nature 369:233–235

Wei H, Perry DC (1996) Dantrolene is cytoprotective in two models of neuronal cell death. J Neurochem 67:2390–2398

White RJ, Reynolds IJ (1995) Mitochondria and Na^+/Ca^{2+} exchange buffer glutamate-induced calcium loads in cultured cortical neurons. J Neurosci 15:1318–1328

White RJ, Reynolds IJ (1996) Mitochondrial depolarization in glutamate-stimulated neurons: an early signal specific to excitotoxin exposure. J Neurosci 16:5688–5697

Wieloch T (1985) Hypoglycemia-induced neuronal damage prevented by an N-methyl-D-aspartate antagonist. Science 230:681–683

Wroblewski JT, Nicoletti F, Costa E (1985) Different coupling of excitatory amino acid receptors with Ca^{2+} channels in primary cultures of cerebellar granule cells. Neuropharmacology 24:919–921

Wyllie AH, Kerr JF, Currie AR (1980) Cell death: the significance of apoptosis. Int Rev Cytol 68:251–306

Xia Z, Dudek H, Miranti CK, Greenberg ME (1996) Calcium influx via the NMDA receptor induces immediate early gene transcription by a MAP kinase/ERK-dependent mechanism. J Neurosci 16:5425–5436

Yang J, Liu X, Bhalla K, Kim CN, Ibrado AM, Cai J, Peng T, Jones DP, Wang X (1997) Prevention of apoptosis by Bcl-2: release of cytochrome c from mitochondria blocked. Science 275:1129–1132

Yuzaki M, Forrest D, Curran T, Connor JA (1996) Selective activation of calcium permeability by aspartate in Purkinje cells. Science 273:1112–1122

Zamzami N, Susin SA, Marchetti P, Hirsch T, Gómez-Monterrey I, Castedo M, Kroemer G (1996) Mitochondrial control of nuclear apoptosis. J Exp Med 183:1533–1544

Editor-in-charge: Professor D. Pette